Origami Inspirations

Great Stellated Dodecahedron-like assembly of the Whipped Cream Star (top) and Truncated Rhombic Triacontahedron (bottom). See pages 69 and 82, respectively.

Origami Inspirations

Meenakshi Mukerji

A K Peters, Ltd.
Natick, Massachusetts

Editorial, Sales, and Customer Service Office

A K Peters, Ltd.
5 Commonwealth Road, Suite 2C
Natick, MA 01760
www.akpeters.com

Library of Congress Cataloging-in-Publication Data

Mukerji, Meenakshi, 1962-
  Origami inspirations / Meenakshi Mukerji.
    p. cm.
  Includes bibliographical references.
  ISBN 978-1-56881-584-8 (alk. paper)
  1. Origami. 2. Polyhedra--Models. I. Title.
  TT870.M823 2010
  736'.982--dc22

                                    2010009411

Printed in India

14 13 12 11 10   10 9 8 7 6 5 4 3 2 1

To Ratul, Rohit, Rohan, Riddhijit, and Rukmini

Contents

Preface

I was certain that I was done after writing my first two books *Marvelous Modular Origami* (2006, A K Peters, Ltd.) and *Ornamental Origami: Exploring 3D Geometric Designs* (2008, A K Peters Ltd.), nearly back to back. But it seemed the more I toyed around with paper, the more this intrigued my imagination, and new modular designs kept emerging. With so many new ideas and such kind readers' encouragement, I easily have enough new creations for yet another book—and here it is.

The first chapter contains origami basics and other material that simply must be included in any modular origami book to be complete. This might be the first book a new folder picks up, as his or her interest just piqued, and certainly all the tools to tackle the projects must be present. If you are already familiar with these basics, please feel free to skip ahead. I have provided polyhedron charts and color distribution charts for referencing during the assembly phase. Folding tips, types of paper, and other material important to folding origami, particularly modular origami, have been included.

The general structure of the book starts with simple models, gradually progressing into more complex ones, although there may be a few exceptions. We begin with simple cubes and then move on to more complicated and interesting models such as the four-sink windmill base models and the decorative dodecahedra folded from pentagons, just to mention a few. The mathematics of these models, of about high school level, is discussed whenever appropriate for those who might enjoy these things, such as myself.

Thanks to the amazing Internet making the entire world available, I am able to interact with like-minded modular origami designers worldwide. Their beautiful creations just further my belief that origami is a language spoken across the globe. In the last chapter, creations of guest artists Aldo Marcell of Nicaragua, Carlos Cabrino (Leroy) of Brazil, Daniel Kwan of the USA, and Tanya Vysochina of Ukraine are included. I have enjoyed and followed their work for some time and I am happy to debut their prolific creations in my book. It was quite a difficult task to decide which of their many fine models to include. I am fortunate they share my philosophy that great origami is to be shared, not kept away in a closet. It is my privilege to introduce them to a new horizon of fellow origami enthusiasts. If you have not come across their work already, you are in for a treat.

The diagrams in this book follow standard origami symbols. Each creator has his or her own diagramming style. The styles of guest contributors have been preserved instead of changing them simply for the sake of being uniform across the book. Sometimes a different viewpoint on folding can quite change the experience of creating a piece. This is a big bonus in presenting a more global selection of models. The origami diagramming language is very powerful. Although there are written descriptions provided with the diagrams most of the time, their need is virtually nonexistent except for certain special circumstances. To keep the diagrams simple, I have refrained from showing layers except when absolutely necessary.

This book covers a range of folding levels from simple to high intermediate models. There is an emphasis toward the latter, which should appeal to audiences with or without a mathematical background, age 12 years and older. While some of the mathematics has been discussed, it is not a requirement to understand the mathematics in order to fold the models. It is just a bonus if you enjoy it, but no knowledge of math is required to produce these stunning objects. You will enjoy the book no matter what your folding level might be, particularly if you are a modular origami lover. Sometimes the units or modules separately may seem quite uninteresting, but I assure you that the assembled finished model is always like a pleasant surprise waiting to be cherished at the end. In modular origami the sum of the whole is almost always more than the sum of the individual parts. And that is what keeps me going and I hope it does the same for you.

Cupertino, California
January 2010

Acknowledgments

It is a pleasurable, although overwhelming task to thank the many people who were so generous with their time and ideas in helping me complete this book. I am certain I will inadvertently miss a few and my apologies in advance for this. I'll begin by thanking those who directly contributed to this book whether by folding models, submitting designs or photographs, or by proofreading and providing valuable feedback. First, a great BIG thanks to guest origami artists Aldo Marcell (Nicaragua), Leroy (Brazil), Daniel Kwan (NJ), and Tanya Vysochina (Ukraine) for their generous design contributions and spectacular photographs. Thanks to Mark Morden (WA) for sharing his novel locking method which I used in Chapter 3. Thanks to Priti Hansia (CA), Tripti Singhal (CA), Halina Narloch (Poland), and Rosalinda Sanchez (AZ), not only for testing my diagrams, but also for making some of my models and providing stunning photographs for the book. Thanks to Rachel Katz (NY) and Jean Jaiswal (OR) for proofreading, diagram testing, and giving me wonderful suggestions. Thanks to Sebastian Janas (Poland), Joy Dutta (CA), and Kedar Amladi (CA) for their stellar photography. Kedar deserves an extra special mention for patiently doing several photo shoots of many of my models. Thanks to Koustubh Oka (CA) for folding a model for the book and generously allowing me to borrow from his reference library. Thanks to Shuzo Fujimoto San and J. C. Nolan for granting me permission to publish photos of models based on their models. Thanks to my editor Charlotte Henderson and my entire publishing team at A K Peters, Ltd. for doing a phenomenal job with the finished book.

Next I'd like to thank those who are indirectly involved with this book by supporting me in many different ways. Thanks to Gaurita and Pradip Amladi, parents of Kedar, and Vinita and Indra Singhal, parents of Tripti, for facilitating their children's contributions to the book. So many people have given me encouragement from so many different parts of the world, it is a bit humbling. I wish to thank the fans of my website, http://www.origamee.net, for their continued support and encouragement (see page 120 for some of the comments left on my guest book). A final note of thanks goes to everyone in my family as well as my friends over here in the US, in India, and around the world for providing me with so much inspiration.

Photo Credits

◆ *Photos by Kedar Amladi*

Cover, page 74: Star with Spirals and close up

Page 26: Flower Cube and Flower Cube 4 Variation

Page 39: Hydrangea Cube and Andrea's Rose Cube

Page 40: Flower Dodecahedra 1 and 2

Page 62: Whipped Cream Star and Windmill Base Cubes

Page 67: Windmill Base Cubes

Back cover, page 72: Whipped Cream Star

Page 80: Three Interlocked Triangular Prisms and Camellia

◆ *Folding and photos by Carlos Cabrino (Leroy)*

Page 93: Chrysanthemum Variation

Page 96: Carnation

◆ *Folding by Priti Hansia and photos by Joy Dutta*

Page 12: Plain Cubes

Page 25: Ray Cube Variation S3, Whirl Cube Variation S3, and Thatch Cube Variation S2

Page 36: Flower Cube 3

◆ *Folding and photos by Daniel Kwan*

Page 83: Truncated Rhombic Triacontahedron

Page 87: Four Interlocked Triangular Prisms

◆ *Folding and photos by Aldo Marcell*

Page 110: Adaptable Dodecahedron,

Page 112: Adaptable Dodecahedron 2

◆ *Photo by Ratul Mukerji*

Page 118: Author photograph

◆ *Folding by Halina Narloch and photos by Sebastian Janas*

Page 52: Flower Dodecahedron and Flower Dodecahedron 2

Page 54: Flower Dodecahedron 3

Page 56: Flower Dodecahedron 4

Page 61: Oleander

◆ *Folding by Koustubh Oka*

Page 34: Flower Cube with Flower Finish A

◆ *Folding and photos by Rosalinda Sanchez*

Page ii: Whipped Cream Star and Truncated Rhombic Triacontahedron

◆ *Folding by Tripti Singhal and photo by Indra Singhal*

Page 73: Star with Spirals

◆ *Folding and photos by Tanya Vysochina*

Page 81: Lily of the Nile

Page 99: Camellia and Variation

Page 101: Dahlia

Page 102: Dahlia Variation

Page 106: Crystal Variation A

All other models are folded and photographed by the author.

1 ◆ Introduction

The word origami is based on two Japanese words: *oru* (to fold) and *kami* (paper). Although this ancient art of paper folding started in Japan and China, origami is now a household word around the world. Most people have probably folded at least a paper boat or an airplane in their lifetime. Origami has evolved immensely in the present times and is much more than a handful of traditional models. Modular origami, origami sculptures, and origami tessellations are but some of the newer forms of the art. The method of designing models has also evolved. While some models are designed the old fashioned way using mostly imagination and by trial and error, others are designed with complex mathematical algorithms using the computer.

Modular origami, as the name implies, involves assembling several, usually identical, modules or units to form one finished model. Generally speaking, glue is not required, but for some models it is recommended for increased longevity, while for some others glue might be essential simply to hold the units together. The models presented in this book do not require glue to stay together, but if you intend to handle the models frequently or to ship them, a bit of glue is a fine idea for only those models that have relatively weaker locks.

While an understanding of mathematics is useful in designing these models, it is not crucial for merely following instructions to construct them. I think that even though mathematics may not be one's strong point, one can still construct these models, and perhaps the process might impart a deeper appreciation for the mathematical principles involved. Like any multi-stepped task it requires patience, diligence, and a bit of practice. It is always a pleasure to see the finished model at the end; the outcome is often greatly different than the individual parts would have initially suggested. Aesthetics and mathematics brilliantly come together in these wonderful modular origami structures to satisfy our many senses.

Modular origami can be fit relatively easily into one's busy schedule if one can be a bit organized. Unlike many other art forms, long uninterrupted stretches of time are not required. This makes it a perfect artistic endeavor given the hectic, fast-paced life we all lead. Upon mastering one unit, which usually does not take long, several more can be folded anywhere anytime, including the short breaks between other chores. When the units are all folded, the final assembly can also be done slowly over time. Modular origami is great for folding during the inevitable waits at airports, doctors' offices, or even on long flights. Just remember to carry your paper, diagrams, and maybe a box for the finished three-dimensional units.

Assembly of the units that comprise a model may at first seem very puzzling to the novice, or may even seem downright impossible. But understanding certain aspects can considerably simplify the process. First, one must determine whether a unit is a face unit, an edge unit, or a vertex unit, that is, whether a unit identifies with a face, an edge, or a vertex, respectively, of a polyhedron. Face units are the easiest to identify. For example, the four-sink windmill base models presented in Chapter 3 are face units. There are only a few known vertex units, e.g., David Mitchell's Electra [Mit00] and Ravi Apte's Universal Vertex Module [Tan02]. Most modular units tend to be edge units. For edge units there is a second step involved—one must identify which part of the unit, which is far from looking like an edge, actually maps to the edge of a polyhedron. Although it may appear perplexing at first, on closer look one may find that it is not an impossible task. Once the identifications are made and the folder can see through the maze of superficial designs and perceive the unit as a face, an edge, or a vertex, assembly becomes simple. It is then just a matter of following the structure of the underlying polyhedron to assemble the units.

Origami Tips

◈ Use paper of the same thickness and texture for all units. This ensures that the finished model's look and strength will be uniform. Virtually any paper from printer paper to giftwrap may be used to fold origami.

◈ Make sure that the grain of the paper is oriented the same way for all your units. To determine the grain of the paper, gently bend paper both horizontally and vertically. The grain of the paper is said to lie along the direction that offers less resistance.

◈ Accuracy is particularly crucial in modular origami, so your folds need to be as accurate as possible. Only then will the finished models look symmetric, neat, and appealing.

◈ It is advisable to fold a trial unit before folding the real units. In some models the finished unit is much smaller than the starting paper size, while in others this is not so. Making a trial unit will give you an idea of what the size of the finished units—and hence a finished model—might be, starting with a certain paper size. It will also give you an idea about the paper properties and whether the paper type selected is suitable for the model you are making.

◈ After you have determined your paper size and type, procure all the paper you need for the model before starting. If you do not have all the paper at the beginning, you may discover, as has been my experience, that you are not able to find more paper of the same kind to finish your model.

◈ If a step looks difficult, looking ahead to the next step often helps immensely. This is because the execution of a current step results in what is diagrammed in the next step.

◈ Assembly aids such as miniature clothespins or paper clips are often advisable, especially for beginners. Some assemblies simply need them whether you are a beginner or not. These pins or clips may be removed as the assembly progresses or upon completion of the model.

◈ During assembly, putting together the last few units, especially the very last one, can be challenging. During those times, remember that it is paper you are working with and not metal! Paper is flexible and can be bent or flexed for ease of assembly.

◈ After completion, hold the model in both hands and compress gently to make sure that all the tabs are placed securely and completely into their corresponding pockets. Finish by working around the ball.

◈ Use templates in unusual folding such as folding into thirds, so as to reduce unwanted creases. The templates in turn can be created using origami methods.

◈ Procure a minimal set of basic handy tools listed next. These tools assist in sizing paper, making neat and crisp creases, curling paper, and assembling models.

Origami Tools

Although origami purists would probably not use any tools to fold, there are many situations during folding that warrant the use of some basic tools. For example, when the parts of the paper to maneuver become too small or unreachable, use of tweezers is quite common.

Creasing Tools. The most basic tool that is used in origami is a bone folder. It allows you to make precise and crisp creases and prevents your nails from being sore when folding excessively. Substitutes might be credit cards or other similar objects.

Cutting Tools. Although cutting is prohibited in pure origami, cutting tools are required for the initial sizing of the paper. A great cutting tool would be a paper guillotine but it is bulky and may not be readily accessible to all people. I find a portable photo trimmer with replaceable blades to be a great substitute. They are inexpensive and easily carried anywhere. Scissors may be used but it is very difficult to get straight cuts.

Curling Tools. Many origami models involve curling. Chop sticks, knitting needles, screwdrivers, or similar objects such as narrow pencils work well for curling paper.

Other Tools. Miniature clothespins may be used during model assembly as temporary aids to hold two adjacent units together. The clothespins may be removed as the assembly progresses or after completion. Tweezers, as mentioned earlier, may be used to access hard to reach places or to fold paper that becomes too small to maneuver with fingers.

Origami can be folded from practically any type of paper. But every model has some paper that works best for it; you will learn which one mostly through experience. Some models might require sturdy paper while some others might require paper that creases softly. The following is a list of commonly available origami paper on the market, as well as references to examples of folded models using the various kinds of paper.

Kami. This is the most readily available origami paper. It is solidly colored on one side and white on the other (see page 25).

Duo. Paper that is one color on one side and a different color on the other (see page 52).

Printer Paper. Paper, white or colored, that is commonly used in home or office computer printers (see page 79).

Mono. These papers have the same color on both sides. They are available in kami weight. Printer paper is another example of mono paper (see page 99).

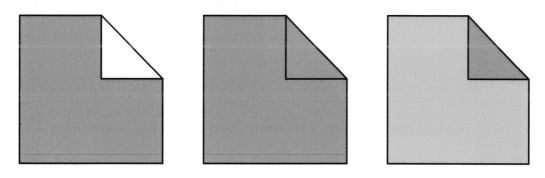

Examples of Kami, Mono, and Duo papers.

Harmony Paper. Paper that has some harmonious pattern formed by various colors blending into one another. They can have dramatic effects on some models (see page 91).

Chiyogami. Origami paper with patterns, usually small, printed on it (see page 80).

Washi. Handmade Japanese paper with plant fiber in the pulp that gives it texture (see page 26 [top model]).

Foil-Backed Paper. These have metallic foil on one side and paper on the other side (see page 32).

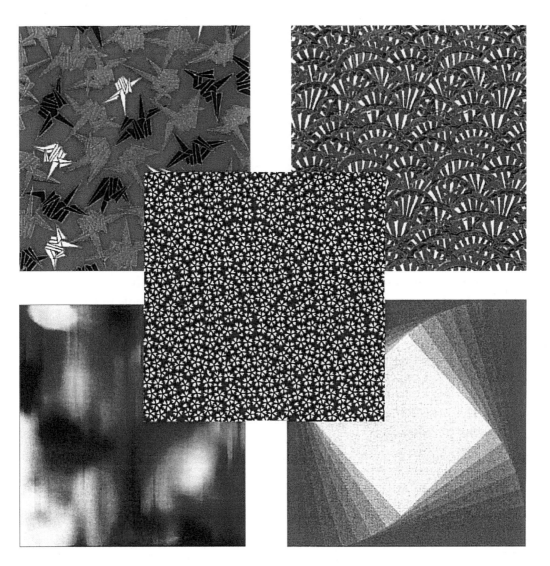

Examples of Washi (top), Chiyogami (middle), and Harmony papers (bottom).

Origami Symbols and Bases

This is a list of commonly used origami symbols and bases.

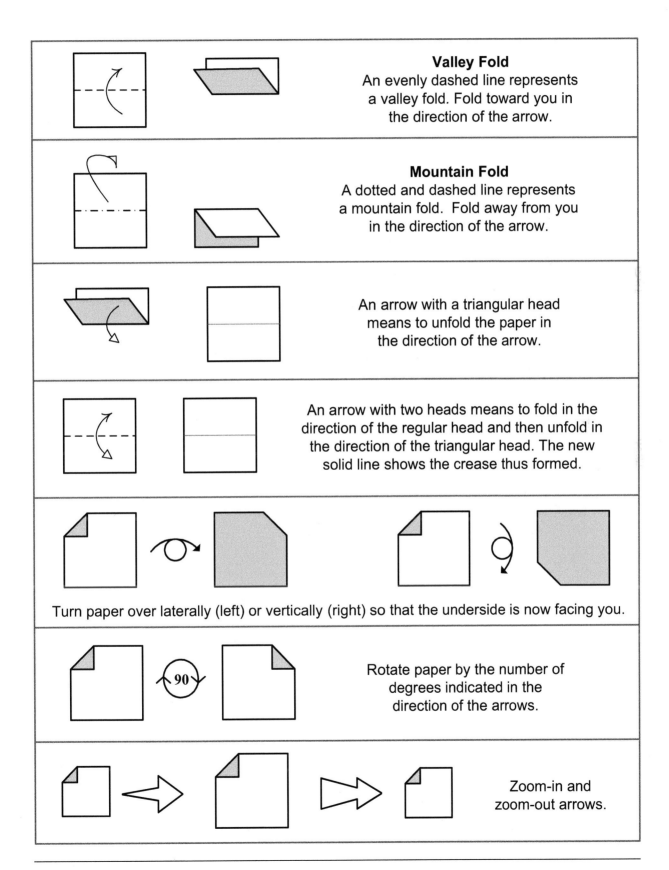

Valley Fold
An evenly dashed line represents a valley fold. Fold toward you in the direction of the arrow.

Mountain Fold
A dotted and dashed line represents a mountain fold. Fold away from you in the direction of the arrow.

An arrow with a triangular head means to unfold the paper in the direction of the arrow.

An arrow with two heads means to fold in the direction of the regular head and then unfold in the direction of the triangular head. The new solid line shows the crease thus formed.

Turn paper over laterally (left) or vertically (right) so that the underside is now facing you.

Rotate paper by the number of degrees indicated in the direction of the arrows.

Zoom-in and zoom-out arrows.

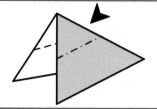 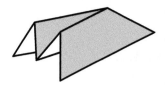

Reverse Fold or Inside Reverse Fold
Push in the direction of the arrow to arrive at the result.

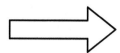 

Pull out paper. Equal lengths. Equal angles.

Figure is truncated for diagramming convenience.

Repeat once, twice, or as many times as indicated by the tail of the arrow.

   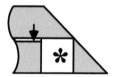 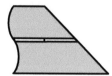

Fold from dot to dot with the circled point as pivot.

✱ : Tuck in opening underneath.

 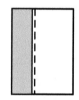

Fold repeatedly to arrive at the result.

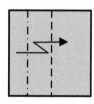 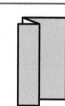 **Pleat Fold**

An alternate mountain and valley fold to form a pleat. Two examples are shown.

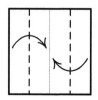

Squash Fold
Turn paper to the right along the valley fold while making the mountain crease such that *A* finally lies on *B*.

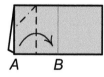 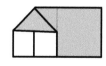

Cupboard Fold
First fold and unfold the centerfold, also called the *book-fold*, then valley fold the left and right edges to the center like cupboard doors.

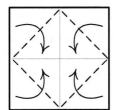  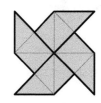

Blintz Base Valley fold and unfold both book-folds. Then valley fold all four corners to the center.

Windmill Base Please see page 64.

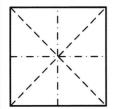 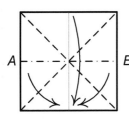 

Waterbomb Base
Valley fold and unfold diagonals, then mountain fold and unfold book-folds. 'Break' line *AB* at the center and collapse such that *A* meets *B*.

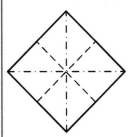 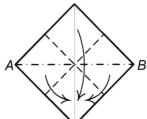

Preliminary Base
This is similar to the waterbomb base above, but the mountain and valley folds are reversed, i.e., the diagonals are mountain folded and the book-folds are valley folded at start.

 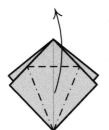  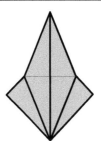

Petal Fold
The figure on the left illustrates petal folding on a flap of a preliminary base.

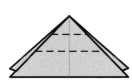 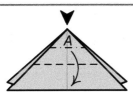 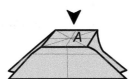 

Spread Squash Figure shows spread squashing the tip of a waterbomb base along the two valley creases shown. Note the final position of tip *A* after the squash.

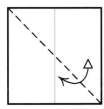 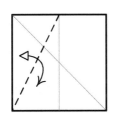 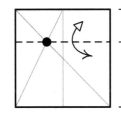 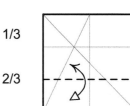

Folding a Square into Thirds Crease book-fold and one diagonal. Crease diagonal of one rectangle to find 1/3 point. Fold and unfold the bottom rectangle into half.

Platonic, Archimedean, and Kepler-Poinsot Solids

The following shows sets of polyhedra commonly referenced during origami constructions.

The Platonic solids, named after the ancient Greek philosopher Plato (428–348 BC), also called the regular solids, are convex polyhedra bound by faces that are congruent regular convex polygons. The same numbers of faces meet at each vertex. There are exactly five.

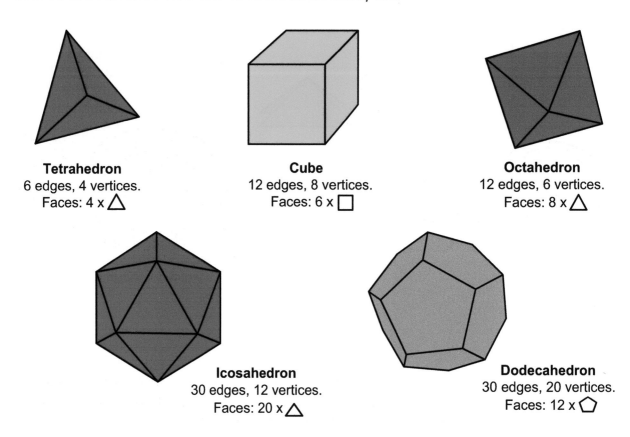

Tetrahedron
6 edges, 4 vertices.
Faces: 4 x △

Cube
12 edges, 8 vertices.
Faces: 6 x □

Octahedron
12 edges, 6 vertices.
Faces: 8 x △

Icosahedron
30 edges, 12 vertices.
Faces: 20 x △

Dodecahedron
30 edges, 20 vertices.
Faces: 12 x ⬠

The Archimedean solids, named after Archimedes (287–212 BC), are semi-regular convex polyhedra bound by two or more types of regular convex polygons meeting in identical vertices. They are distinct from the Platonic solids, which are composed of a single type of polygon meeting at identical vertices. Shown here are eight of the thirteen Archimedean solids. Those not shown are the Snub Dodecahedron, Truncated Tetrahedron, Truncated Dodecahedron, Great Rhombicosidodecahedron, and Great Rhombicuboctahedron.

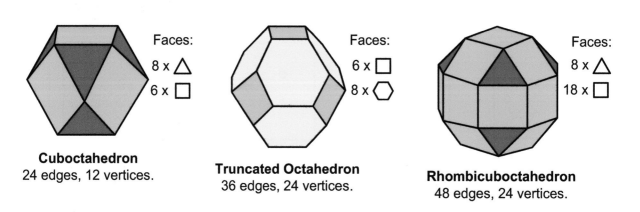

Faces:
8 x △
6 x □

Faces:
6 x □
8 x ⬡

Faces:
8 x △
18 x □

Cuboctahedron
24 edges, 12 vertices.

Truncated Octahedron
36 edges, 24 vertices.

Rhombicuboctahedron
48 edges, 24 vertices.

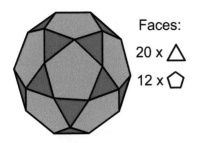

Faces:

20 x △

12 x ⬠

Icosidodecahedron
60 edges, 30 vertices.

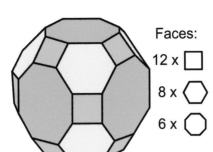

Faces:

12 x ☐

8 x ⬡

6 x ⯃

Truncated Cuboctahedron
72 edges, 48 vertices.

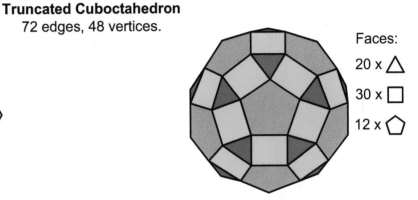

Faces:

32 x △

6 x ☐

Snub Cube
60 edges, 24 vertices.

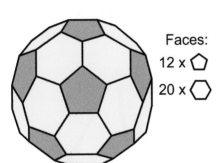

Faces:

12 x ⬠

20 x ⬡

Truncated Icosahedron
90 edges, 60 vertices.

Faces:

20 x △

30 x ☐

12 x ⬠

Rhombicosidodecahedron
120 edges, 60 vertices.

The Kepler-Poinsot solids, named after Johannes Kepler and Louis Poinsot (17th–19th centuries), are four regular concave polyhedra with intersecting facial planes. These can be obtained by stellating Platonic solids. Only two of the four are shown below. The two not shown are the Great Dodecahedron and the Great Icosahedron.

Greater Stellated Dodecahedron

Lesser Stellated Dodecahedron

Even Color Distribution

This section illustrates a few polyhedra with even color distribution for their edges. For most modular origami constructions, each edge of a polyhedron maps to a module or unit.

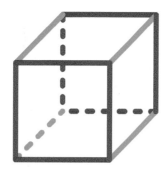

Three-color tiling of a cube
(every vertex has three distinct colors)

Four-color tiling of a cube
(every face has four distinct colors)

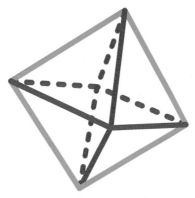

Three-color tiling of an octahedron
(every face has three distinct colors)

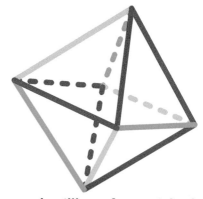

Four-color tiling of an octahedron
(every vertex has four distinct colors)

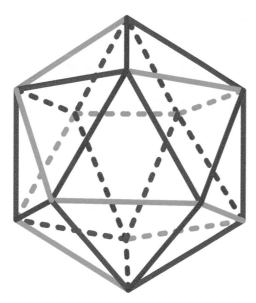

Three-color tiling of an icosahedron
(every face has three distinct colors)

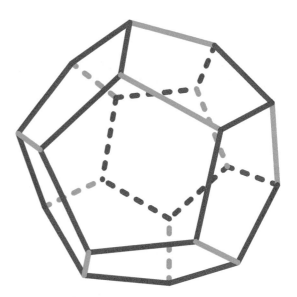

Three-color tiling of a dodecahedron
(every vertex has three distinct colors)

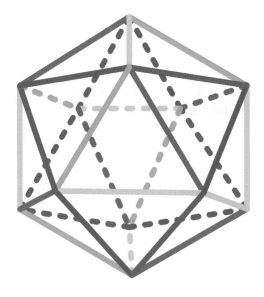

Five-color tiling of an icosahedron
(every vertex has five distinct colors and
every face has three distinct colors)

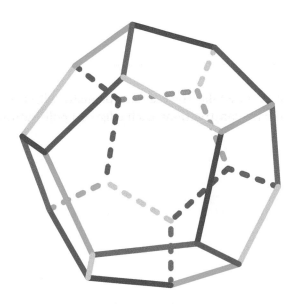

Five-color tiling of a dodecahedron
(every face has five distinct colors and
every vertex has three distinct colors)

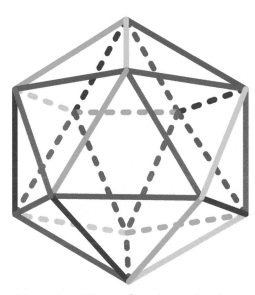

Six-color tiling of an icosahedron
(every vertex has five distinct colors and
every face has three distinct colors)

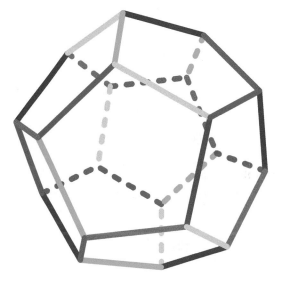

Six-color tiling of a dodecahedron
(every face has five distinct colors and
every vertex has three distinct colors)

For color to be distributed evenly, the number of colors one chooses must be a factor of the number of edges or units in a model. For example, for a 30-unit model, one can use three, five, six, or ten colors. In all of these figures, each edge represents one origami unit. The dotted lines are edges invisible from the point of view. The number of units you need to fold per color is equal to the total number of edges divided by the total number of colors. Even color distribution makes a model more appealing but for some models the use of a single color may be more effective while, for some others random coloring works fine. You can try semi-even coloring as well. For example, say we select four colors for a 30-unit model— yellow, blue, red, and green with six units of yellow, blue, and green each, but with twelve units of red. We could tile the red as if it were covering for both red and pink in the top two figures and the rest exactly as shown. The result is pleasing as well.

Top: Plain Cube and Plain Cube 2. Middle: Ray Cube and Thatch Cube.
Bottom: Whirl Cube and Thatch Cube in reverse coloring.

2 ◆ Simple Cubes

The cube is one of the simplest polyhedral shapes but nevertheless it is excellent for stepping into the polyhedral origami world for the first time. In this chapter we will make cubes consisting of 6, 12, or 24 units, with an emphasis on the latter. They all utilize the technique of folding a square into thirds to obtain pockets on two opposite edges of the units. This method is widely used in a host of modular models, including Robert Neale's Penultimate [Pla] series of models and David Mitchell's Simplex Cubes [Mit].

The 24-unit cubes have patterns made by exposing the white side of kami paper. Revealing the white side to create patterns is often called "color change" in the origami world. Any model that involves color change is also potentially a great candidate for folding with *duo* paper. Pattern variations have been discussed for the 24-unit cubes. For these cubes, it is fascinating to see how merely switching tab and pocket orientations opens up the possibility of numerous different patterns. Also note that if you start with a square of side of length *a*, then the edges of the finished cubes will be of length $2a/3$.

Below is shown a traditional method of folding a square into thirds. You can fold just one square as a template using the same size paper as your model units. Then use the template for folding your other squares into thirds to eliminate unwanted crease lines as well as to save time.

Folding a Square into Thirds

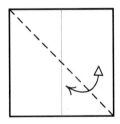

1.

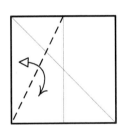

2.

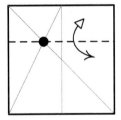

3.

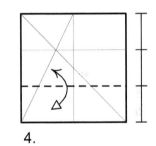

4.

Proof that the above method generates thirds

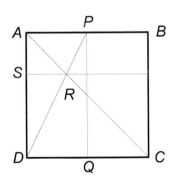

Let the side length of starting square *ABCD* be *a*. Let *PQ* be the line along which the square is folded into half. Let *R* be the point where the two diagonal creases meet and *SR* be the creased horizontal line extended to edge *BC*.

$\angle CAD = 45°$ because *AC* is the diagonal of the square. Hence $SR = AS \dots$ (i)

Now, right $\triangle PAD$ and right $\triangle RSD$ are similar because they share a common angle $\angle ADP$.

Hence, $AD/SD = AP/SR$,

or $a/SD = (a/2)/SR$,

or $DS/SR = 2$,

or $DS = 2SR = 2AS$ using (i) above.

Therefore, *DS* is twice *AS*, which means that *AS* is 1/3 of *AD*.

Plain Cube

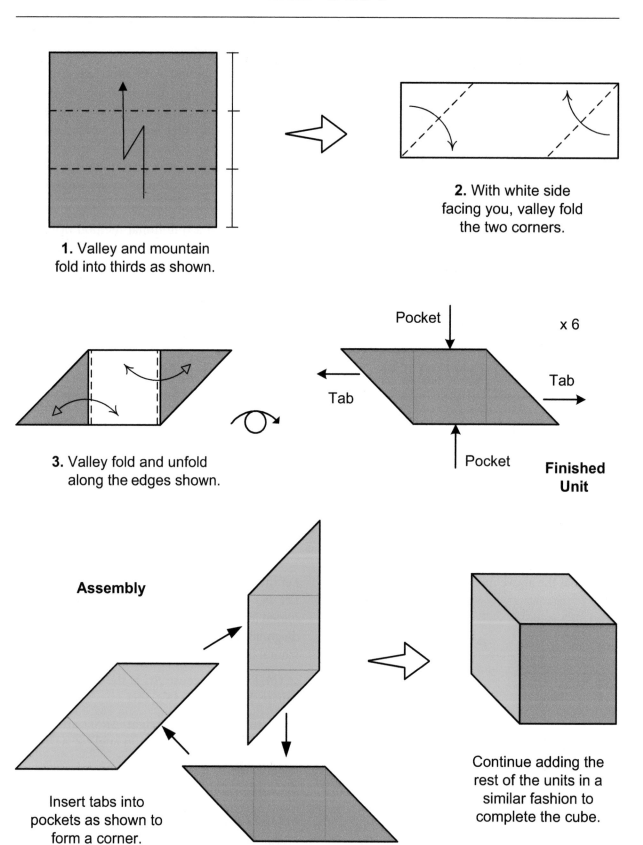

1. Valley and mountain fold into thirds as shown.

2. With white side facing you, valley fold the two corners.

3. Valley fold and unfold along the edges shown.

Pocket

x 6

Tab

Tab

Pocket

Finished Unit

Assembly

Insert tabs into pockets as shown to form a corner.

Continue adding the rest of the units in a similar fashion to complete the cube.

See page 12 for finished model photo.

Plain Cube 2

Do Steps 1 and 2 of Plain Cube on the previous page and then unfold the folds of Step 2.

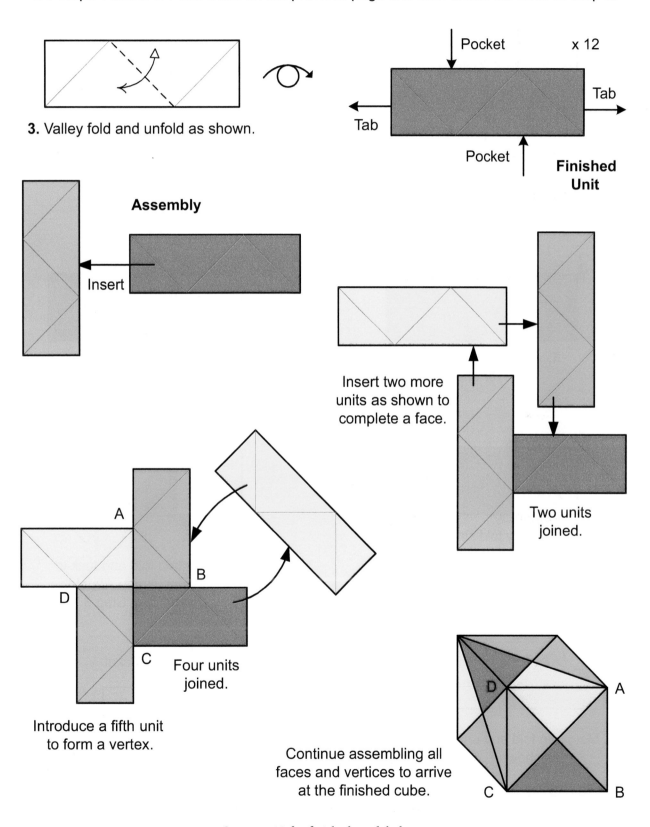

3. Valley fold and unfold as shown.

Pocket   x 12

Tab

Tab

Pocket

Finished Unit

Assembly

Insert

Insert two more units as shown to complete a face.

Two units joined.

A

B

D

C

Four units joined.

Introduce a fifth unit to form a vertex.

Continue assembling all faces and vertices to arrive at the finished cube.

D   A

C   B

See page 12 for finished model photo.

Ray Cube

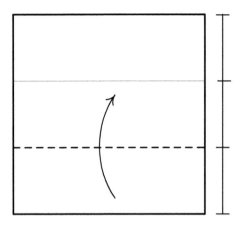

1. Fold square into thirds and unfold. Then valley fold bottom crease.

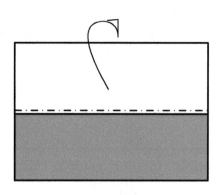

2. Mountain fold along edge.

3. Valley fold corner, top layer only.

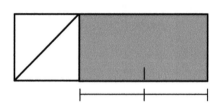

4. Pinch halfway point as shown.

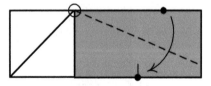

5. Valley fold single layer only to bring top edge to pinched point.

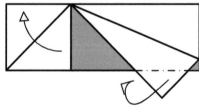

6. Mountain fold and tuck in between layers. Unfold crease from Step 3.

7. Mountain fold and unfold as shown.

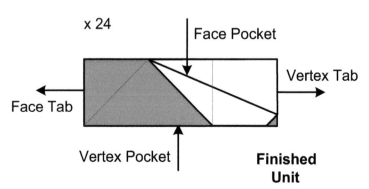

x 24

Face Pocket

Vertex Tab

Face Tab

Vertex Pocket

Finished Unit

Assembly

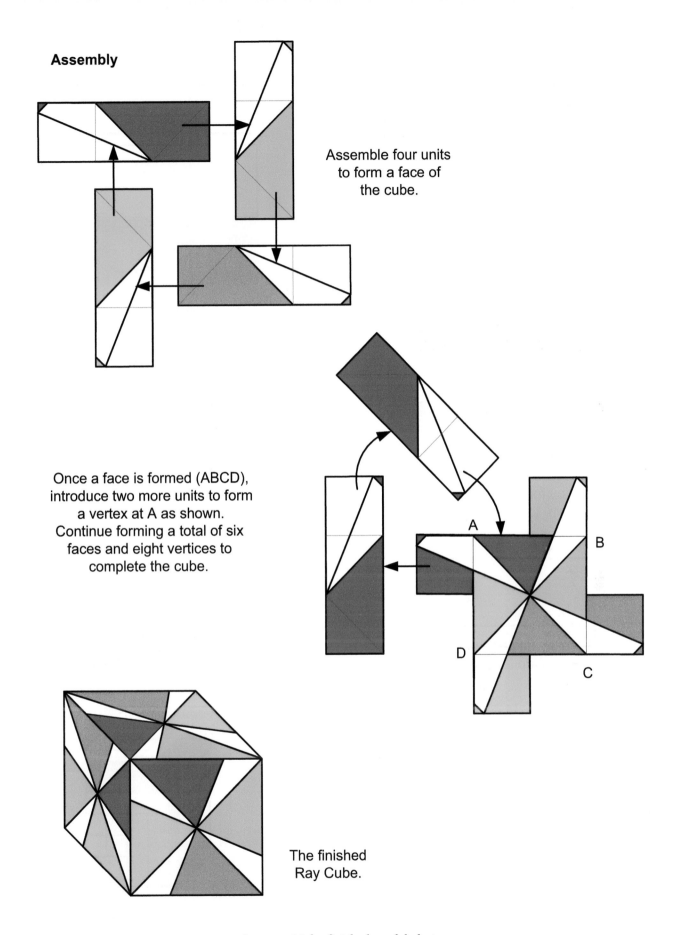

Assemble four units
to form a face of
the cube.

Once a face is formed (ABCD),
introduce two more units to form
a vertex at A as shown.
Continue forming a total of six
faces and eight vertices to
complete the cube.

A

B

D

C

The finished
Ray Cube.

See page 12 for finished model photo.

Thatch Cube

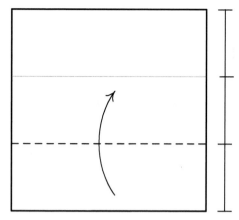

1. Fold square into thirds and unfold. Valley fold bottom crease.

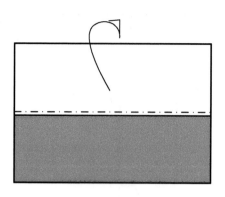

2. Mountain fold along edge.

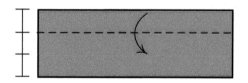

3. Valley fold about 1/3 way down, top layer only.

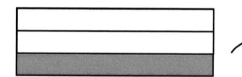

4. Turn over.

5. Valley fold and unfold corner, top layer only.

6. Valley fold and unfold as shown.

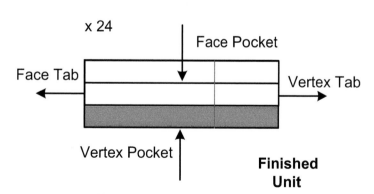

x 24

Face Pocket

Face Tab

Vertex Tab

Vertex Pocket

Finished Unit

Assembly: Assemble exactly as explained in Ray Cube.

Finished Cube.

See page 12 for finished model photo.

Whirl Cube

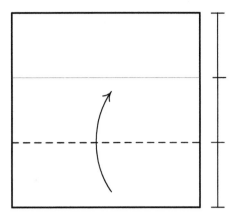

1. Fold square into thirds and unfold. Valley fold bottom crease.

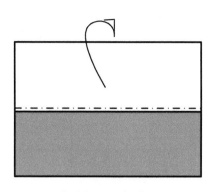

2. Mountain fold along edge.

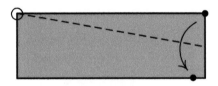

3. Valley fold to bring corner to edge, top layer only.

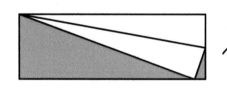

4. Turn over.

5. Valley fold and unfold corner, top layer only.

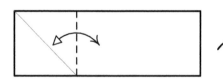

6. Valley fold and unfold as shown.

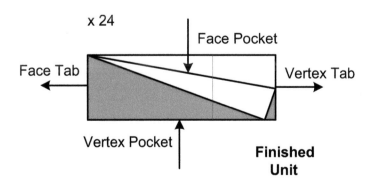

x 24

Face Pocket

Face Tab

Vertex Tab

Vertex Pocket

Finished Unit

Assembly: Assemble exactly as explained in Ray Cube.

Finished Cube.

See page 12 for finished model photo.

Pattern Variations

Interesting variations of the Ray Cube, Thatch Cube, and Whirl Cube models can be obtained by simply switching the face tab with the vertex tab or the face pocket with the vertex pocket or by switching both. Let's take the Ray Cube as an example and examine the various cases of interchanged tabs and pockets and the resultant face patterns. Let's call the original pattern S0 and the switched patterns S1, S2, and S3.

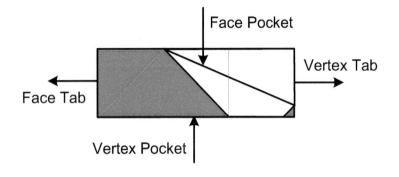

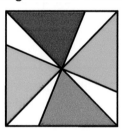

S0: Original unit as diagrammed earlier.

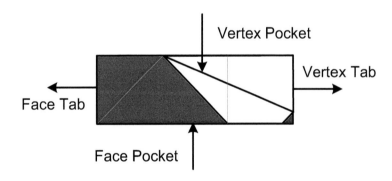

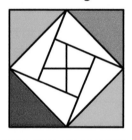

S1: Face and vertex tabs interchanged.

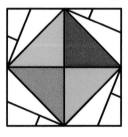

S2: Face and vertex pockets interchanged.

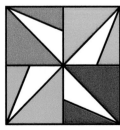

S3: Both pockets and tabs interchanged.

Ray Cube variations, in various four-color schemes, shown both in regular and reverse colorings. Reverse colorings are indicated with (R).

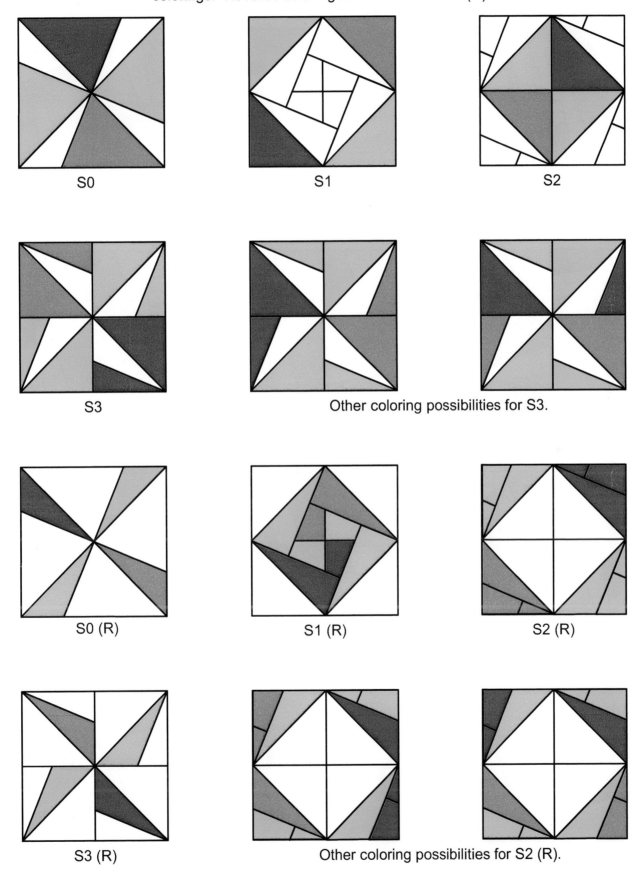

S0

S1

S2

S3

Other coloring possibilities for S3.

S0 (R)

S1 (R)

S2 (R)

S3 (R)

Other coloring possibilities for S2 (R).

Whirl Cube variations, in various four-color schemes, shown both in regular and reverse colorings. Reverse colorings are indicated with (R).

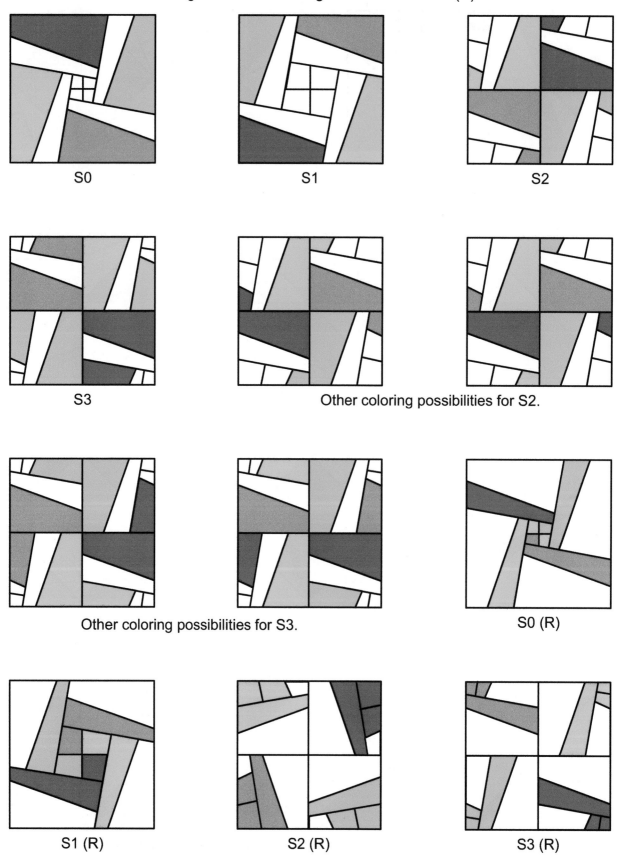

S0

S1

S2

S3

Other coloring possibilities for S2.

Other coloring possibilities for S3.

S0 (R)

S1 (R)

S2 (R)

S3 (R)

Thatch Cube variations, in various four-color schemes, shown both in regular and reverse coloring. Variations that are merely mirror images of another variation have been omitted.

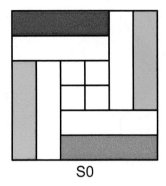

S0

S1 is mirror image of S0.
S3 is mirror image of S2.

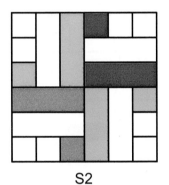

S2

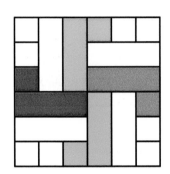 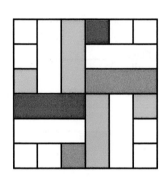

Other coloring possibilities for S2.

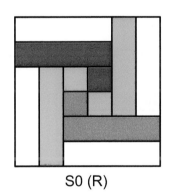

S0 (R)

S1 (R) is mirror image of S0 (R).
S3 (R) is mirror image of S2 (R).

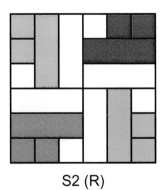

S2 (R)

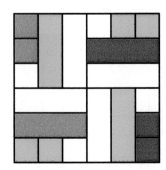 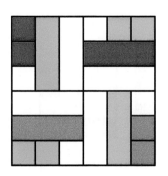

Other coloring possibilities for S2 (R).

Ray Cube, Whirl Cube, and Thatch Cube variations shown with units made of same colored kami. Using the same color for all units accentuates the designs formed by color change.

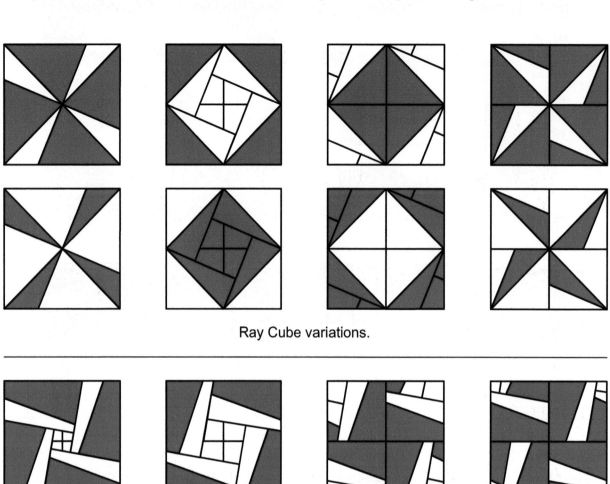

Ray Cube variations.

Whirl Cube variations.

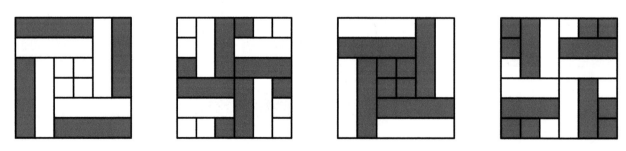

Thatch Cube variations.

Pattern Variations

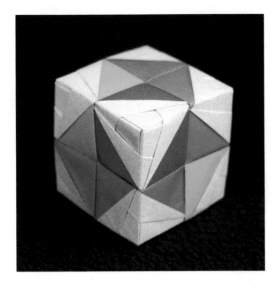
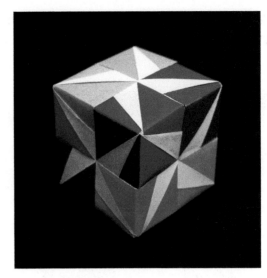

Ray Cube variations S2 and S3 (top). Whirl Cube variations S2 and S3 (middle). Thatch Cube variation S2 in single color (bottom).

Flower Cube with Flower Finish B (top) and Flower Cube 4 variation (bottom).

3 ◆ Four-Sink Base Models

The four-sink windmill base is a wonderful foundation with endless possibilities. While the base is essentially a windmill base with its four corners sunk, the best approach to folding it is not necessarily to follow that order. Discussed first in this chapter is one of the many ways to fold the base. The first model that I came across that uses a predecessor of this base (shown in Step 11 in the next section) is the Loop Kusudama by Saburo Kase [NOA94]. Other models that use the same predecessor are the Rose Faced Cuboctahedron by Tomoko Fuse [Fus92] and a Kusudama by Yasuko Suyama [Ori 00]. Tomoko Fuse's book *Kusudama Origami* [Fus02] also has several models based on it. Flower Bloom Modular and a few other models by Mark Morden start with the four-sink base. Mark's joining method is novel and very sturdy. I have adopted his method of joining the units in my models in this chapter.

The models Flower Cube and Flower Cube 2 resemble J. C. Nolan's Andrea's Rose [Nol95], which is essentially the same but has more levels of fractal tessellations. I discovered this fact soon after creating my models. The resemblance is purely coincidental. Nolan himself found out after creating his rose that his model was in fact a slightly varied version of Toshikazu Kawasaki's Pinecone [Kas98]. Nolan says in his book *Creative Origami*, "At first I was surprised and alarmed, but I have since learned that it is common for the same model to be created by two different persons independently of each other, especially if they are structurally simple." I completely agree with his statement. Others, including Nick Robinson, have discovered this model independently as well. It is worthwhile mentioning that six units of any level of Andrea's Rose may be joined by the method used in this chapter to form a cube. Similarly, six units of Shuzo Fujimoto's Hydrangea [Ada] may also be connected to form a cube. Photos of the latter two models are shown at the end of the chapter but the models are not diagrammed.

Recommendations

Paper Size: 6"–8" squares

Paper Type: Any

Finished Model Size

Cubes with approximately 2"–2.5" sides

Four-Sink Base

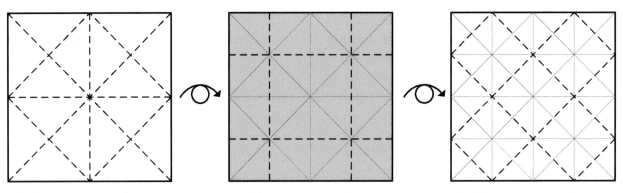

1. Precrease as shown.

2. Precrease as shown.

3. Precrease as shown.

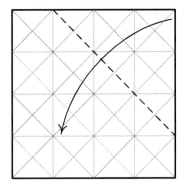

4. Valley fold along pre-existing crease.

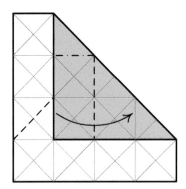

5. Valley and mountain fold to arrive at result.

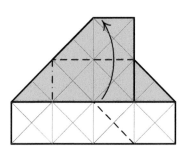

6. Valley and mountain fold to arrive at result.

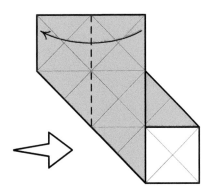

7. Valley fold to lift paper partway. Do not flatten.

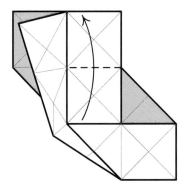

8. Valley fold partway two layers. Do not flatten.

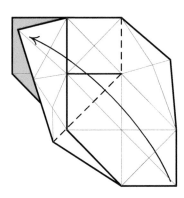

9. Valley fold as shown to bring corners together.

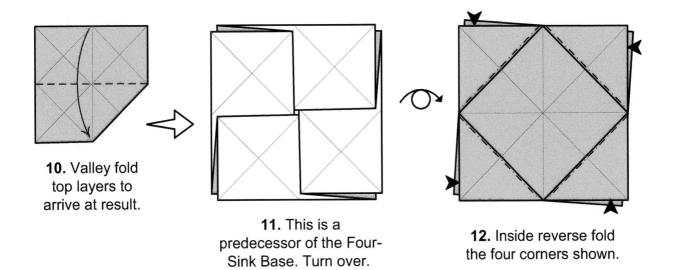

10. Valley fold top layers to arrive at result.

11. This is a predecessor of the Four-Sink Base. Turn over.

12. Inside reverse fold the four corners shown.

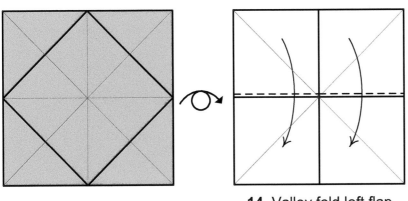
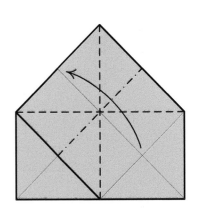

13. Turn over.

14. Valley fold left flap followed by the right.

15. Collapse top flap like a waterbomb base.

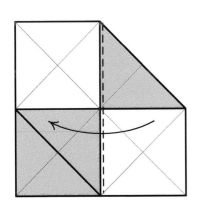
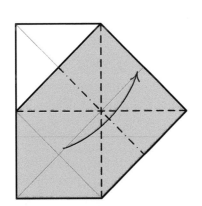
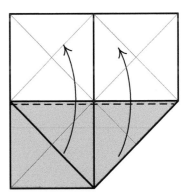

16. Valley fold top flap.

17. Collapse top flap like a waterbomb base.

18. Valley fold left flap and then right flap.

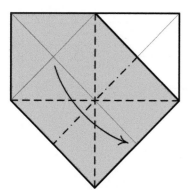

19. Collapse top flap
like a waterbomb base.

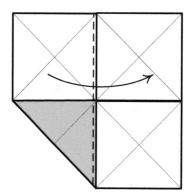

20. Valley fold top flap.

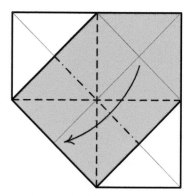

21. Collapse top flap
like a waterbomb base.

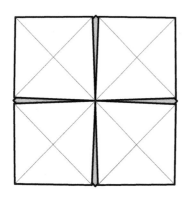

The completed
Four-Sink Base.

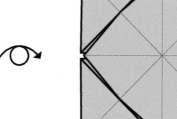

The front view of
the base.

Butterfly Cube

Make a Four-Sink Base as explained in the previous section. To avoid creases at the center from showing, do not crease the center of the paper when doing Step 1 of the base.

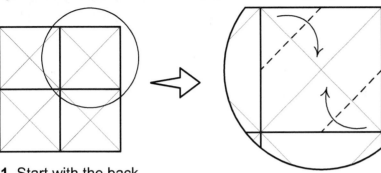

1. Start with the back side of the Four-Sink Base facing you.

2. Fold the two corners shown.

3. Fold the corner in.

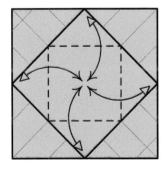

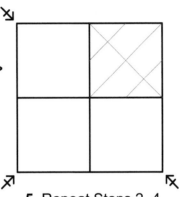

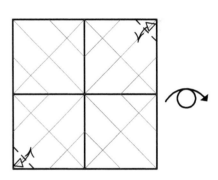

4. Unfold to Step 1.

5. Repeat Steps 2–4 at other corners.

6. Fold and unfold tips slightly to mark tabs (optional). Turn over.

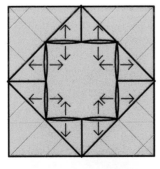

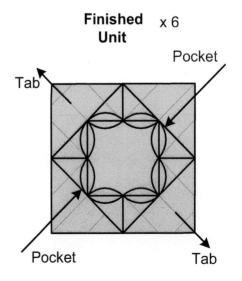

7. Fold and unfold the four corners shown, then orient them to point towards you.

8. For each corner now pointing towards you, spread open the slits and curve outwards.

Finished Unit x 6

Tab

Pocket

Pocket

Tab

Assembly

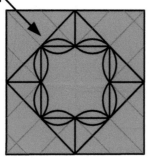

Insert tab
into pocket.

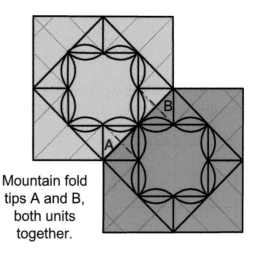

Mountain fold
tips A and B,
both units
together.

Mountain fold line PQ, both units
together. PQ will lie along one
edge of the finished cube. Continue
assembling all six units similarly
such that each unit lies on a face of
the finished cube and each locked
joint lies on an edge of the cube.

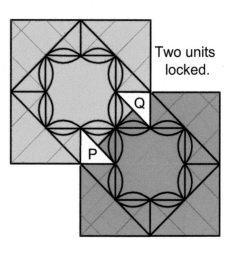

Two units
locked.

Butterfly Cube.

Flower Cube

This Flower Cube can be made with three different flower finishes. They have been named Finishes A, B, and C. Start by doing Steps 1–7 of the previous model, the Butterfly Cube, as diagrammed on page 31.

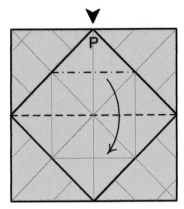

8. Spread squash the corner marked P. The corner P will end up at the center of the unit.

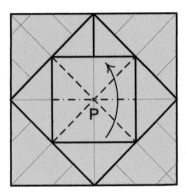

9. Perform a waterbomb fold on the top square at P as shown. The result is essentially sinking P.

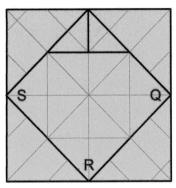

10. Repeat Steps 8 and 9 on corners Q, R, and S to sink them to the center.

The next step will be different for the three flower finishes A, B, and C as shown below.

Flower Finish A

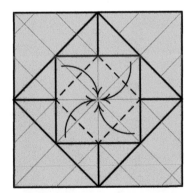

11. Valley fold the four corners to the center. The corners will naturally unfold slightly.

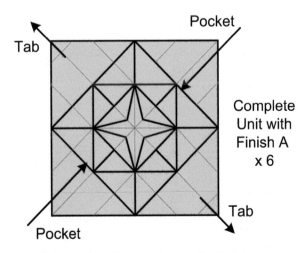

Complete Unit with Finish A x 6

Assemble like the Butterfly Cube.

Flower Finish B

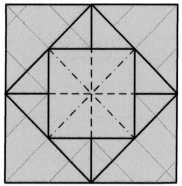

11. Mountain and valley fold the center square as shown, so its corners stand up to arrive at Flower Finish B.

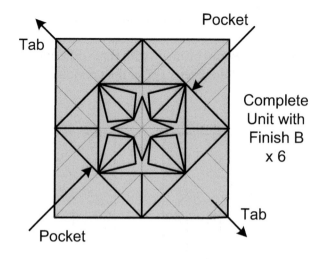

Tab

Pocket

Complete Unit with Finish B x 6

Tab

Pocket

Flower Finish C

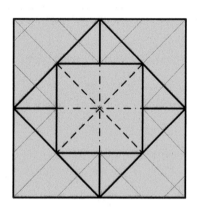

11. Reverse the mountain and valley folds of Step 11 above.

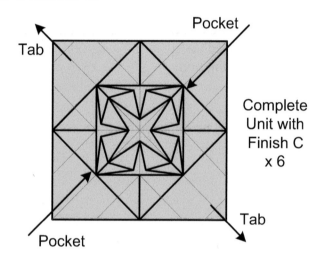

Tab

Pocket

Complete Unit with Finish C x 6

Tab

Pocket

Flower Cube with Flower Finishes A (left) and B (right). See also page 26.

Flower Cube 2

Start by doing up to Step 10 of the previous model, the Flower Cube, on page 33.

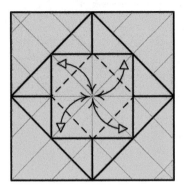

11. Valley fold and unfold
the four corners shown.

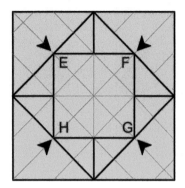

12. Perform Steps 8–10 of the Flower
Cube model to spread squash and
sink the corners E, F, G, and H.

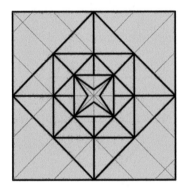
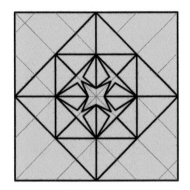

Completed units showing flower finishes A, B, and C. Make six
of the kind you want to construct and join like the Butterfly Cube.

Flower Cube 2 with Flower Finishes B (left) and C (right).

Flower Cube 3

Start by doing Steps 1–7 of the Butterfly Cube, as diagrammed on page 31.

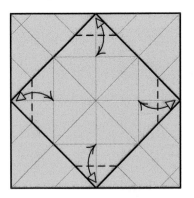

8. Valley fold and unfold the four corners shown.

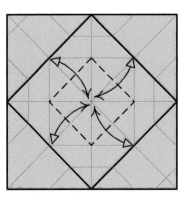

9. Valley fold and unfold top square only to crease an inner square.

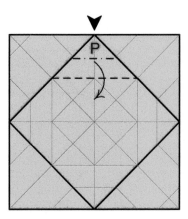

10. Spread squash the corner marked P up to the valley fold mark.

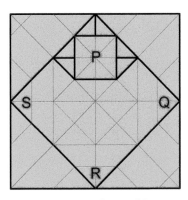

11. Repeat Step 10 on corners Q, R, and S.

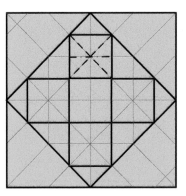

12. Mountain and valley fold as shown to raise slightly like petals.

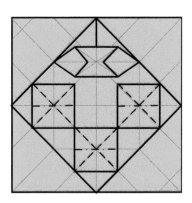

13. Repeat Step 12 on all other squares.

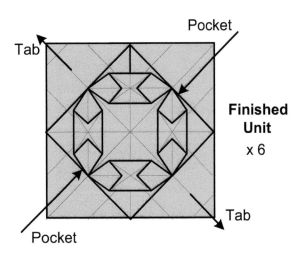

Finished Unit
x 6

Pocket

Tab

Tab

Pocket

Assemble like the Butterfly Cube.

Flower Cube 4

Start by completing up to Step 11 of the previous model, the Flower Cube 3, on page 36.

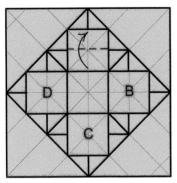

12. Valley fold the small square into half.

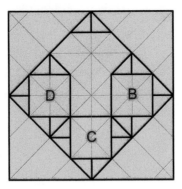

13. Repeat Step 12 on squares B, C, and D.

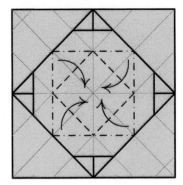

14. Mountain and valley fold as shown. The squares A, B, C, and D should reappear on top after this step.

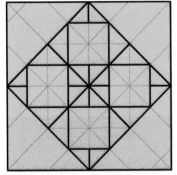

15. Lift and round the edges of the innermost square outwards with your nails.

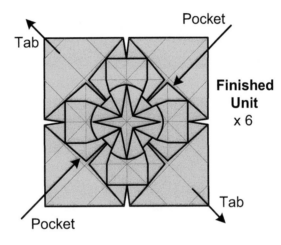

Tab

Pocket

Finished Unit x 6

Pocket

Tab

Variation

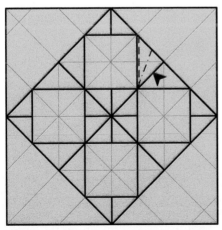

From Step 15 above, squash a corner as shown before rounding edges. Repeat on all seven remaining corners.

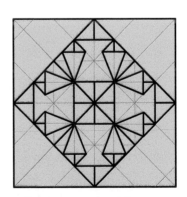

Tuck squashed flaps under the squares below them and then round edges like Step 15 above.

Flower Cube 4 (top) and its variation (bottom). See also page 26.

Shuzo Fujimoto's four-level Hydrangeas [Ada] (top) and J. C. Nolan's four-level Andrea's Roses [Nol95] (bottom) locked into cubes using the same method as the other models in this chapter. The units are not diagrammed in this book but the photographs are shown to encourage readers to find the diagrams and to try these enjoyable models.

Flower Dodecahedron (top) and Flower Dodecahedron 2 (bottom).

4 ◆ Folding with Pentagons

Origami almost always means folding with a square piece of paper. But folding with other regular polygons such as triangles, pentagons, hexagons, or octagons is not unheard of. While squares are readily available, one has to go the extra mile to obtain the other polygons. These polygons can, of course, be made by various origami methods. Some examples of folding with polygons are Rona Gurkewitz and Bennett Arnstein's Dodecahedron Flower Ball [Gur95] made with triangles, Philip Shen's Ten Pointed Star [Jac89] made from a pentagon, Roberto Gretter's Flower With A Star [Bos03] made from a hexagon, and Paul Jackson's Octagonal Flower [Jac89] made from an octagon. Almost all the models in Gurkewitz's and Arnstein's book *Multimodular Origami Polyhedra: Archimedeans, Buckyballs and Duality* [Gur03] start with various polygons other than squares.

In this chapter we will explore some models that begin with pentagons. While playing with a pentagonal piece of paper in the late 1990s, I mapped the folds of a traditional Lily/Iris from a square piece of paper onto a pentagon. The resulting flower was more beautiful and lifelike and I was quite pleased. Although there are many origami artists who have tried this pentagonal version of the traditional model, i.e., the idea is not new, I will be presenting the traditional lily and its pentagonal five-point version in this chapter for the sake of demonstrating how the folds of a square piece of paper can be mapped/extrapolated/translated to a pentagonal piece of paper. The exercise will hopefully encourage you to take other simple models, such as traditional stars or flowers made from squares, and map them to pentagons. Some examples of models that map well from square to pentagon are the Saar Star by Endla Saar [Gro05] and my Tuberose [Muk08] and Oleander (page 60) models, which are essentially square and pentagonal versions, respectively, of the same model.

There are several techniques for obtaining a pentagon starting with a square piece of paper. Diagrammed in this chapter is a method by Toyoaki Kawai [Kaw70] that is widely used. I find the technique easy and elegant with minimal wastage of paper. The resulting pentagon is close to perfect with minimal error and it works very well for normal scale origami purposes.

Recommendations

Paper Size: 4.5"–5" squares

Paper Type: Paper heavier than kami

Finished Model Size

Approximately 4.5"–5" in height

Making a Pentagon and Its Mathematics

Shown below is a traditional method of making a pentagon from a square by Toyoaki Kawai [Kaw70].

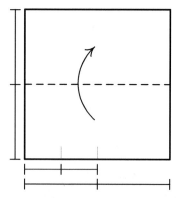

1. Pinch two half points on bottom edge, then fold in half.

2. Pinch half point on right edge.

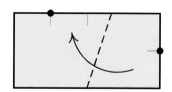

3. Valley fold to match dots.

4. Valley fold to match edges.

5. Unfold to Step 2.

6. Crease to drop a perpendicular from corner to new line. Refold to Step 5.

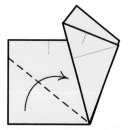

7. Valley fold to match edges.

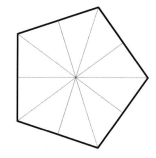

8. Mountain fold.

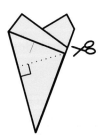

9. Cut all layers along crease made in Step 6.

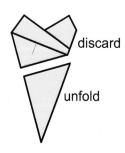

discard

unfold

10. Discard top piece and unfold bottom to arrive at pentagon.

The resulting pentagon will have side $a \sin 36°$ if the starting square is of side length of a.

Shown below is the proof that the pentagon obtained by the Kawai method, described in the previous page, is close enough to a regular pentagon. Also calculated is the side length of the pentagon in terms of the side length of the starting square $ABCD$. Let the side length AB be a.

This figure shows the layout of the unfolded pentagon along with the creases made up to Step 6. EF is the fold from Step 1 and P, Q, and R are the half-point pinches made in Steps 1 and 2. Red lines show additional construction required for the proof.

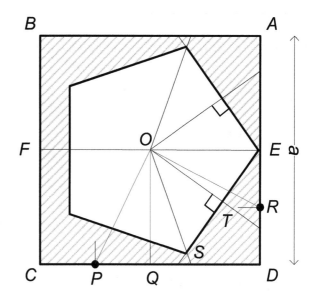

In right $\triangle REO$,

$\tan \angle EOR = ER/OE = (a/4)/(a/2)$,

so $\angle EOR = \tan^{-1} 0.5 = 26.57°$.

OS bisects $\angle POR$ by virtue of the fold in Step 3, hence $\angle POS = \angle SOR$.

Let $\angle POS = \angle SOR = \theta$.

In right $\triangle OPQ$, $\tan \angle POQ = PQ/OQ = (a/4)/(a/2)$, so $\angle POQ = \tan^{-1} 0.5 = 26.57°$.

$\angle POF = $ right $\angle QOF - \angle POQ = 90 - 26.57 = 63.43°$.

Now, straight $\angle FOE = \angle POF + \angle POS + \angle SOR + \angle EOR$.

Substituting the values, $180 = 63.43 + \theta + \theta + 26.57$. Therefore, $\theta = 45°$.

Then, $\angle EOS = \angle EOR + \angle ROS = 26.57 + 45 = 71.57° \approx 72°$.

But $\triangle EOS$ is one of the five identical sections of the pentagon and $\angle EOS \approx 72°$, i.e., 360°/5, which is a property of a regular pentagon. Also, all sides of the pentagon are identical by virtue of the rest of the folding sequence and cutting. Hence it is a very close approximation to a regular pentagon.

Calculating the side length:

In right $\triangle EOT$, $\angle EOT = 72/2 = 36°$ and $OE = a/2$.

Hence $TE = a/2 \sin 36°$.

Therefore, side length of the pentagon = $2 \times a/2 \times \sin 36° = a \sin 36°$.

Traditional and Five-Point Lilies (see page 45).

Making the Joint Units

Here is how to make the joint units, which will be used for four of the models in this chapter, the Flower Dodecahedra 1–4. Start by folding Steps 1–3 of making a pentagon as discussed on page 42.

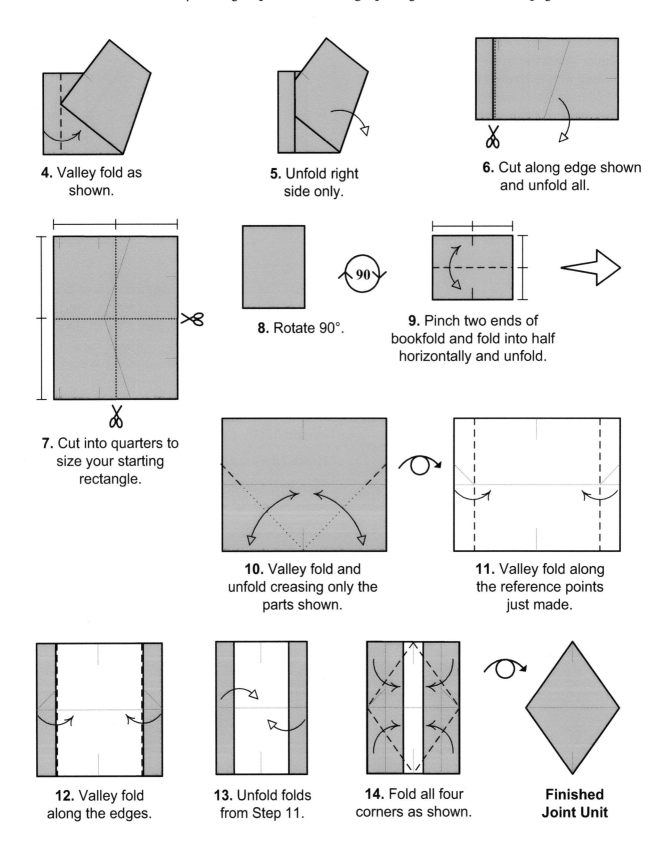

4. Valley fold as shown.

5. Unfold right side only.

6. Cut along edge shown and unfold all.

7. Cut into quarters to size your starting rectangle.

8. Rotate 90°.

9. Pinch two ends of bookfold and fold into half horizontally and unfold.

10. Valley fold and unfold creasing only the parts shown.

11. Valley fold along the reference points just made.

12. Valley fold along the edges.

13. Unfold folds from Step 11.

14. Fold all four corners as shown.

Finished Joint Unit

Traditional and Five-Point Lilies

The purpose of presenting this model is to demonstrate how to map the folds of a square piece of paper to a pentagonal one. Shown below are the steps for making a Traditional Lily from a square sheet of paper, which most of you probably already know, alongside the steps for making a Five-Point Lily from a pentagonal sheet of paper. It is assumed that the reader has folded a Traditional Lily in the past, so not every detail is given in the diagrams.

On studying the first step of a Traditional Lily, you will see that to make the waterbomb base we collapse by valley folding the lines joining the center of the square to the vertices and mountain folding the lines joining the center of the square to the midpoints of the sides. On applying the same logic to a pentagon, we end up with a pentagonal waterbomb base. The rest of the folds are the same, except that for certain steps the folds are repeated four times for the Traditional Lily and five times for the Five-Point Lily.

Traditional Lily

Five-Point Lily

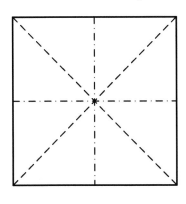

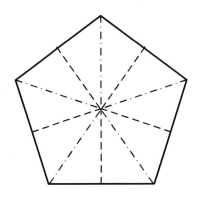

1. Mountain and valley fold as shown to collapse into waterbomb base.

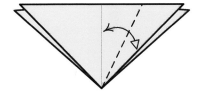

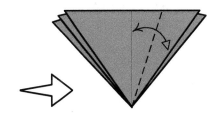

2. Valley fold and unfold top flap only.

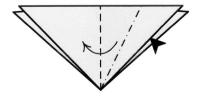

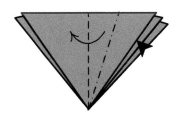

3. Squash the flap you just folded.

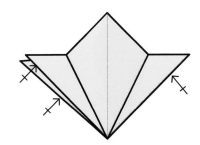

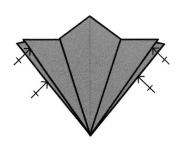

4. Repeat Steps 2 and 3 on all other flaps.

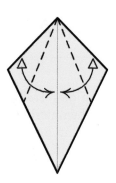 

5. Valley fold and unfold
top flap only.

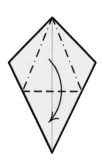 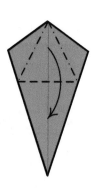

6. Petal fold along the
creases you just made.

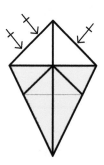 

7. Repeat Steps 5 and
6 on all other flaps.

8. Valley fold top flap.

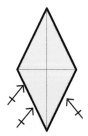 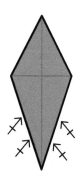

9. Repeat Step 8 on all
other flaps.

Traditional and Five-Point Lilies

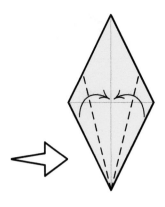

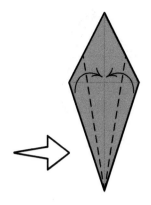

10. Valley fold edges of top flap to center.

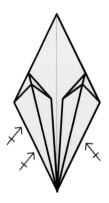

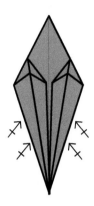

11. Repeat Step 10 on all other flaps.

12. Curl front petal towards you. Repeat on all other petals to arrive at finished Lily.

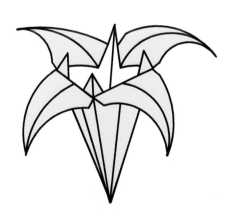

Finished Traditional (left) and Five-Point (right) Lilies.

See page 43 for photo of finished model.

Flower Dodecahedron

Start by making 12 pentagons as explained on page 42. The first few steps involve making a pentagonal windmill base, which is made by mapping the folds of a regular windmill base.

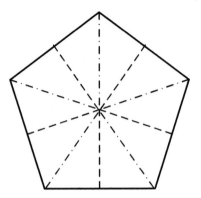

1. Mountain and valley fold along pre-existing creases as shown.

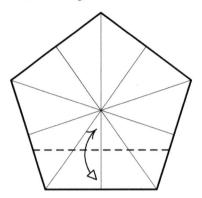

2. Bring bottom edge to center and crease and unfold.

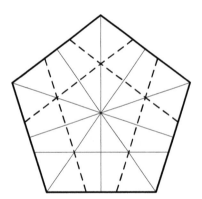

3. Repeat Step 2 on all other edges.

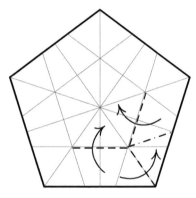

4. Valley and mountain fold as shown.

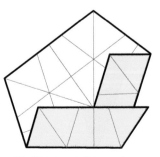

5. Repeat Step 4 on all other corners.

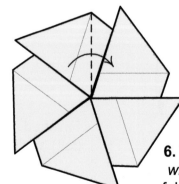

6. This is a *pentagonal windmill base.* Valley fold top flap to the right.

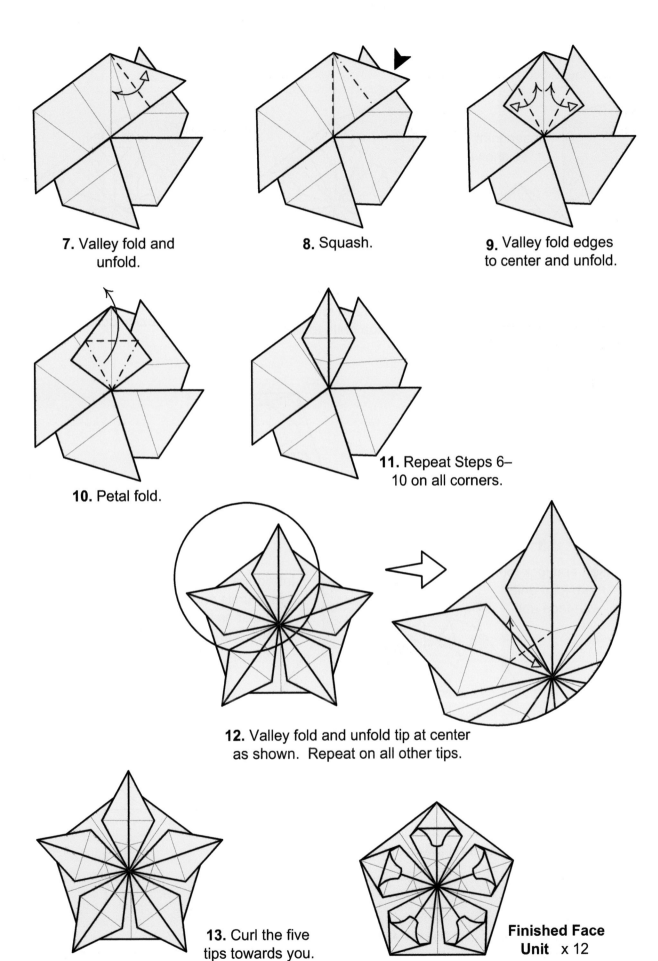

7. Valley fold and unfold.

8. Squash.

9. Valley fold edges to center and unfold.

10. Petal fold.

11. Repeat Steps 6–10 on all corners.

12. Valley fold and unfold tip at center as shown. Repeat on all other tips.

13. Curl the five tips towards you.

Finished Face Unit x 12

Make 30 joint units as explained on page 44 starting with the same-sized square as the face units. When you assemble, the face units will lie on the 12 faces and the joint units will lie along the 30 edges of your finished dodecahedron.

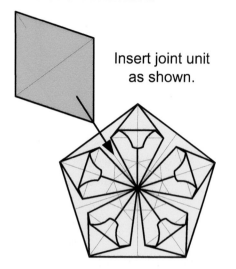

Insert joint unit as shown.

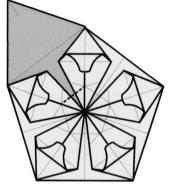

Re-crease valley fold of Step 12 to lock joint and face units together.

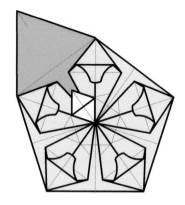

Face and joint units locked.

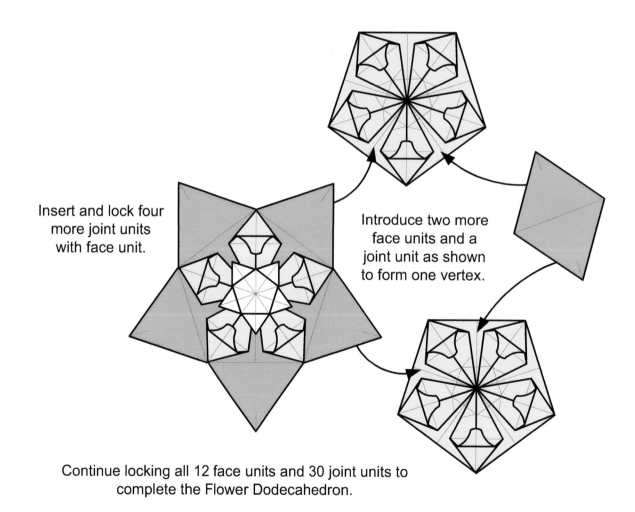

Insert and lock four more joint units with face unit.

Introduce two more face units and a joint unit as shown to form one vertex.

Continue locking all 12 face units and 30 joint units to complete the Flower Dodecahedron.

See pages 40 and 52 for finished model photos.

Flower Dodecahedron

Flower Dodecahedron 2

Start by folding Steps 1–11 of Flower Dodecahedron on page 48.

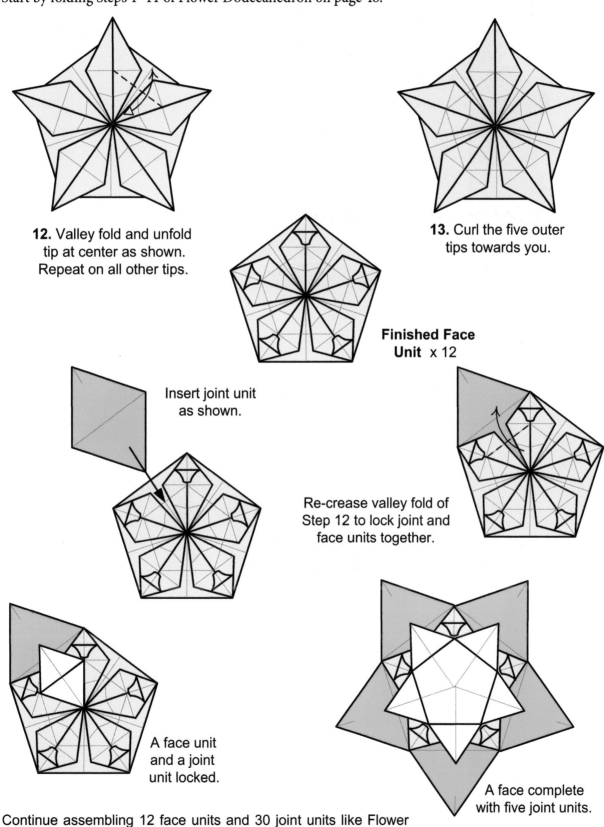

12. Valley fold and unfold tip at center as shown. Repeat on all other tips.

13. Curl the five outer tips towards you.

Finished Face Unit x 12

Insert joint unit as shown.

Re-crease valley fold of Step 12 to lock joint and face units together.

A face unit and a joint unit locked.

A face complete with five joint units.

Continue assembling 12 face units and 30 joint units like Flower Dodecahedron 1 to arrive at the finished Flower Dodecahedron 2.

Flower Dodecahedron made with duo paper (top) and Flower Dodecahedron 2 made with duo paper (bottom).

Start by folding Steps 1–9 of Flower Dodecahedron on page 48.

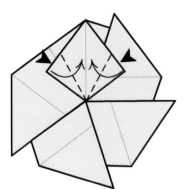

10. Squash fold both sides.

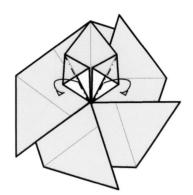

11. Mountain fold to tuck on both sides.

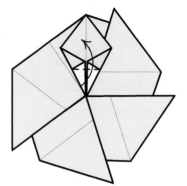

12. Valley fold tip upwards as shown.

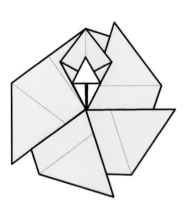

13. Repeat Steps 6–12 on all corners.

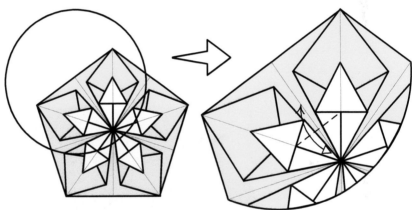

14. Valley fold and unfold tip at center. Repeat on all corners.

Finished Face Unit x 12

Assembly

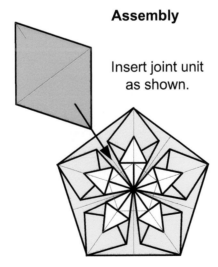

Insert joint unit as shown.

Re-crease valley fold of Step 14 to lock joint and face units together.

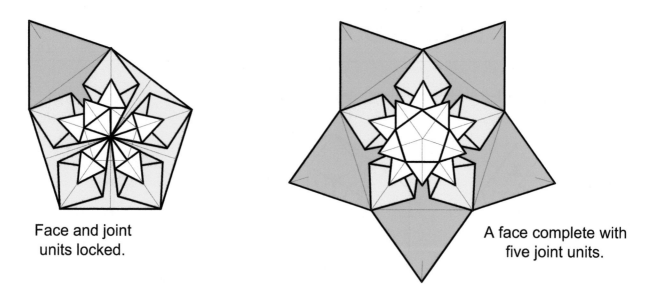

Face and joint
units locked.

A face complete with
five joint units.

Continue assembling 12 face units and 30 joint units like Flower
Dodecahedron 1 to arrive at the finished Flower Dodecahedron 3.

Flower Dodecahedron 3 made with duo paper. See also page 57.

Flower Dodecahedron 4

Start by folding Steps 1–9 of Flower Dodecahedron on page 48.

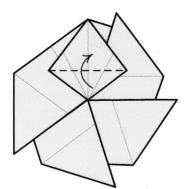

10. Valley fold and unfold as shown.

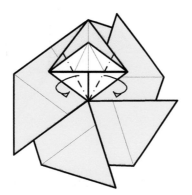

11. Mountain fold to tuck in both sides.

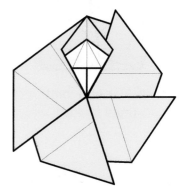

12. Repeat Steps 6 to 11 on all other flaps.

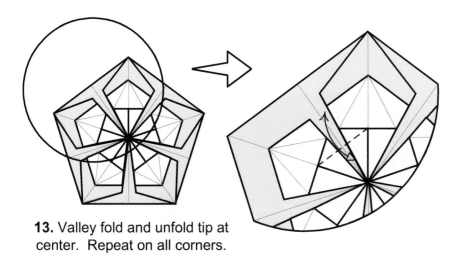

13. Valley fold and unfold tip at center. Repeat on all corners.

Finished Face Unit x 12

Assembly

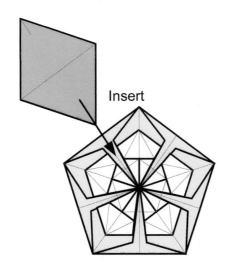

Insert

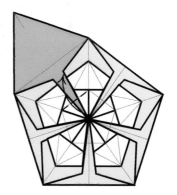

Re-crease valley fold of Step 13 to lock joint and face units together.

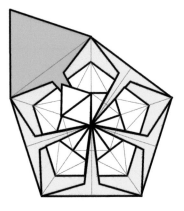

Face and joint
units locked.

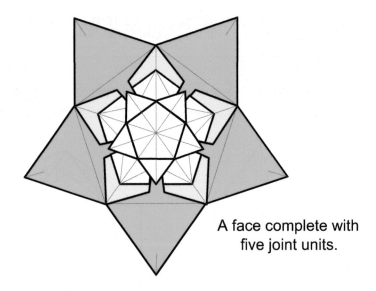

A face complete with
five joint units.

Continue assembling 12 face units and 30 joint units like Flower
Dodecahedron 1 to arrive at the finished Flower Dodecahedron 4.

Flower Dodecahedron 4 made with duo paper. See also page 57.

More examples of Flower Dodecahedron 3 (top) and 4 (bottom).

Flower Dodecahedron 5

Start by folding Steps 1–8 of Flower Dodecahedron on page 48.

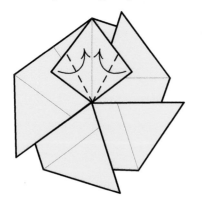

9. Valley fold edges to center.

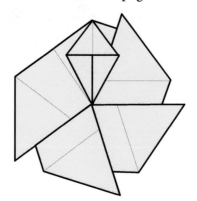

10. Repeat Steps 6–10 on all other corners.

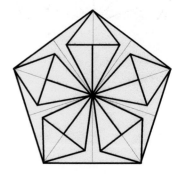

Finished Face Unit x 12

Adjustment to Joint Unit

Make a joint unit as diagrammed on page 44 and then make the following adjustments.

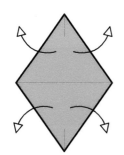

1. Unfold the four corners.

2. Valley fold along existing crease.

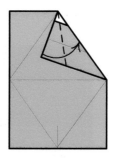

3. Valley fold as shown.

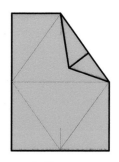

4. Repeat Steps 1–3 on all other corners.

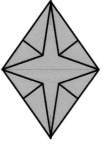

Finished Adjusted Joint Unit x 30

Locking

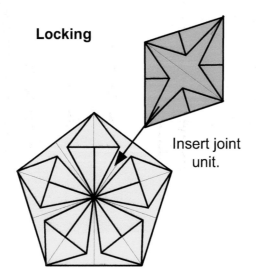

Insert joint unit.

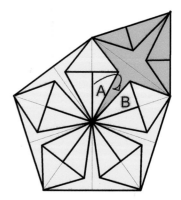

Unfold flap A and tuck it under flap of joint unit. Repeat with flap B.

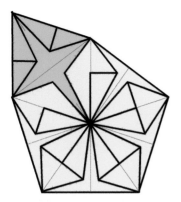

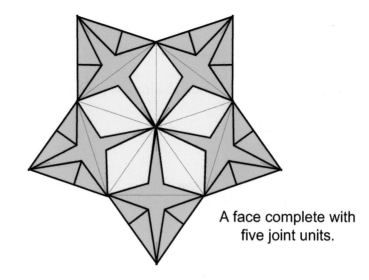

A face unit and a
joint unit locked.

A face complete with
five joint units.

Continue assembling 12 face units and 30 joint units like Flower
Dodecahedron 1 to arrive at the finished Flower Dodecahedron 5.

Flower Dodecahedron 5.

Oleander

Do Steps 1–11 of Flower Dodecahedron on page 48.

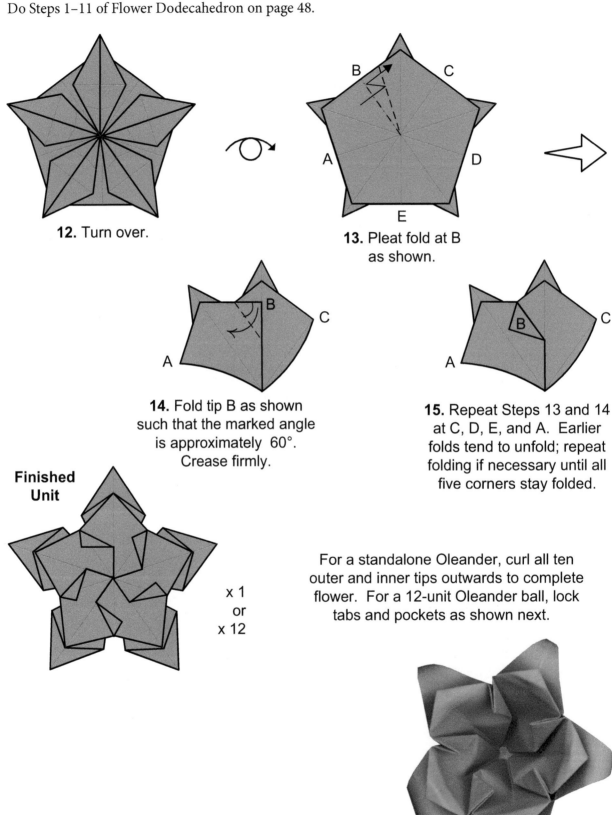

12. Turn over.

13. Pleat fold at B as shown.

14. Fold tip B as shown such that the marked angle is approximately 60°. Crease firmly.

15. Repeat Steps 13 and 14 at C, D, E, and A. Earlier folds tend to unfold; repeat folding if necessary until all five corners stay folded.

Finished Unit

x 1
or
x 12

For a standalone Oleander, curl all ten outer and inner tips outwards to complete flower. For a 12-unit Oleander ball, lock tabs and pockets as shown next.

A Standalone Oleander

Assembly

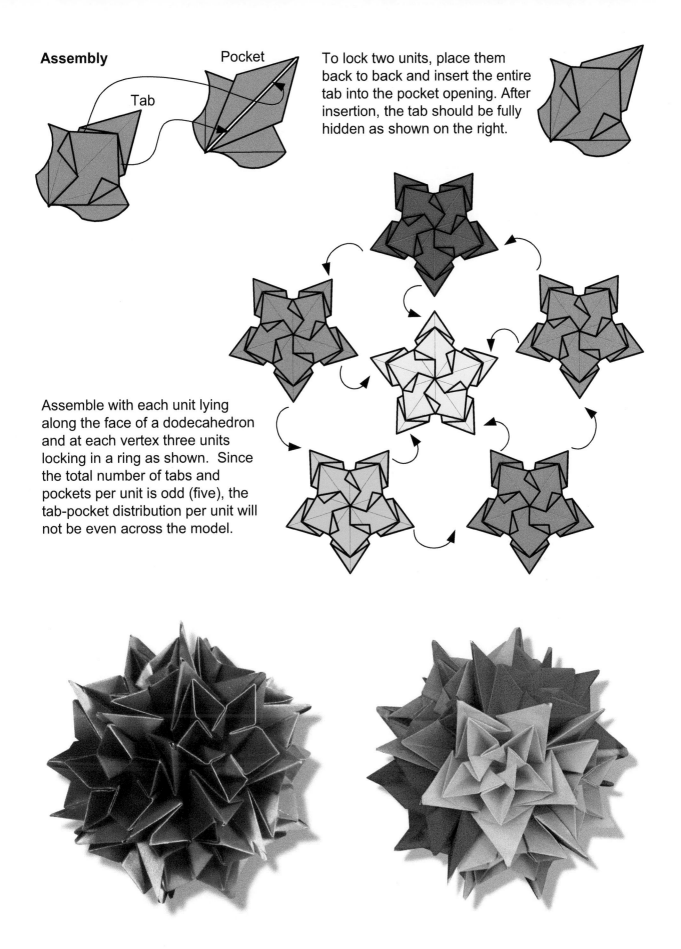

Pocket

Tab

To lock two units, place them back to back and insert the entire tab into the pocket opening. After insertion, the tab should be fully hidden as shown on the right.

Assemble with each unit lying along the face of a dodecahedron and at each vertex three units locking in a ring as shown. Since the total number of tabs and pockets per unit is odd (five), the tab-pocket distribution per unit will not be even across the model.

Oleanders shown in single and six colors.

Whipped Cream Star (top) and Windmill Base Cube and Cube 2 (bottom).

5 ◆ Miscellaneous

In this chapter we will explore two main kinds of models—windmill base cubes and a series I have named Whipped Cream models and their variations. We will also make a Waves model, which is similar to a model in my previous book, *Ornamental Origami* [Muk08], but the units are a one-piece rendition of that earlier model.

The windmill base cubes are perhaps the simplest of the models presented in this book. They are rudimentary enough that it is highly likely that other creators may also have designed them independently. The windmill base, also often referred to as the pinwheel base, is very versatile and numerous models can be folded from it. The traditional Chrysanthemum Kusudama starts with a windmill base. The earliest documented windmill base models were created by Friedrich Froebel in the 1800s. Tomoko Fuse

in her book *Kusudama Origami* [Fus02] has many models developed from the traditional Chrysanthemum, which in turn uses the windmill base. Kunihiko Kasahara has a host of modular models that begin with the windmill base in his book *Extreme Origami* [Kas03]. In my previous book *Ornamental Origami: Exploring 3D Geometric Designs* [Muk08], I have an entire chapter devoted to windmill base models.

The Whipped Cream series of models have been named thus simply for the lack of a better name. When I made my first model in various shades of pink, it reminded me of cake decorations and hence the name. I have shown the models mostly in dodecahedral/icosahedral assembly consisting of 30 units but other polyhedral shapes can be created with 6, 12, or 24 units. Going higher than 30 is possible, but I have not tried.

Recommendations for Windmill Base Models

Paper Size: 4"–6" squares will yield cubes of side 2"–3"

Paper Type: Kami or any other paper

Recommendations for Whipped Cream Models

Paper Size: 4" × 2" rectangles will result in finished model size of 5"–7" in height depending on model

Paper Type: Kami or paper heavier than kami, such as Tant or printer paper

Windmill Base Cube

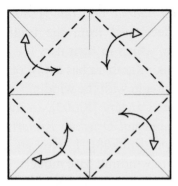 

1. Valley fold the ends of both diagonals and book-folds and unfold. Then fold corners to center and unfold.

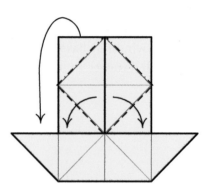

2. Cupboard fold and unfold.

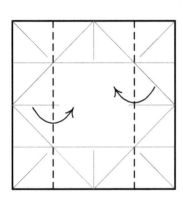

3. Cupboard fold.

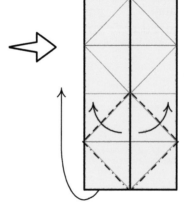

4. Make the mountain and valley folds as shown while bringing bottom edge to middle.

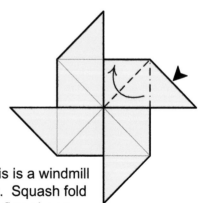

5. Repeat Step 4 on the top, bringing top edge to middle.

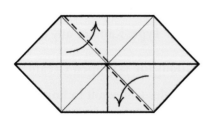

6. Valley fold the two flaps in the directions indicated.

7. This is a windmill base. Squash fold the flap shown.

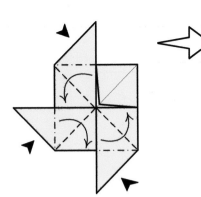

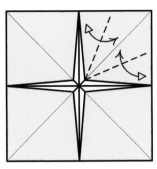

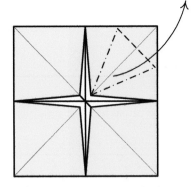

8. Squash fold the other three flaps.

9. Valley fold edges to diagonal and unfold.

10. Valley and mountain fold taking the corner outwards. This is also called *petal fold*.

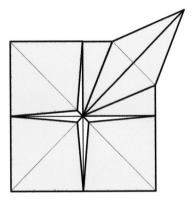

11. Repeat Steps 9 and 10 on the other three corners.

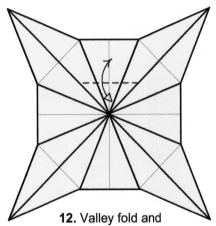

12. Valley fold and unfold approximately below halfway point from top edge to center.

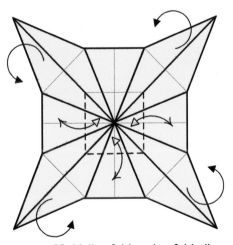

13. Valley fold and unfold all other corners at the same level to form an inner square. Curl all outer tips towards you.

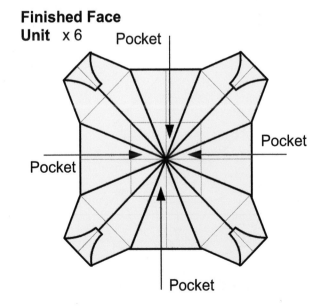

Finished Face Unit x 6

Pocket

Pocket

Pocket

Pocket

Pocket

Making the Joint Units

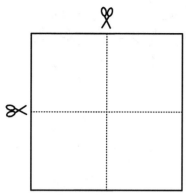

1. Start with same size paper as the face units and cut into quarters.

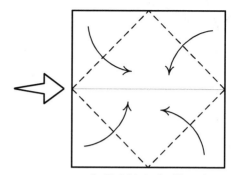

2. Fold into half and unfold. Blintz corners.

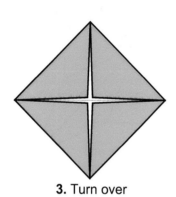

3. Turn over

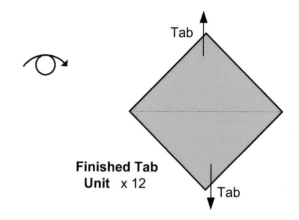

Tab

Tab

Finished Tab Unit x 12

Assembly

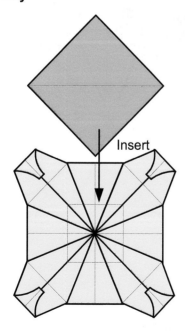

Insert

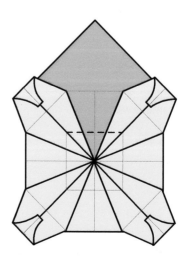

Valley fold face and joint units together at pre-existing crease to lock them together.

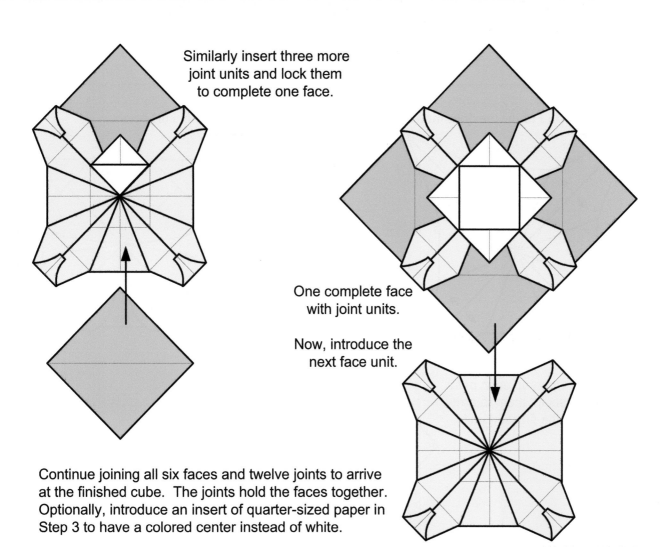

Similarly insert three more joint units and lock them to complete one face.

One complete face with joint units.

Now, introduce the next face unit.

Continue joining all six faces and twelve joints to arrive at the finished cube. The joints hold the faces together. Optionally, introduce an insert of quarter-sized paper in Step 3 to have a colored center instead of white.

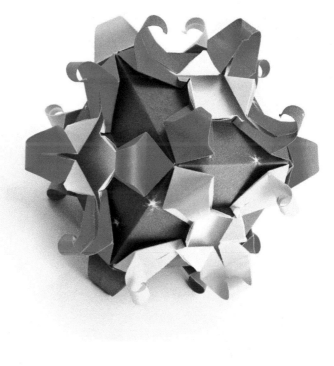

Windmill Base Cube with (left) and without (right) inserts.

Start with Step 12 of the previous model (page 64), color side reversed. Do not curl petals.

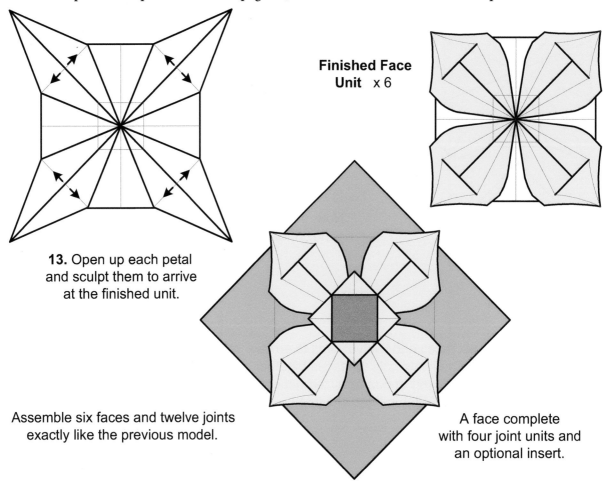

Finished Face Unit x 6

13. Open up each petal and sculpt them to arrive at the finished unit.

Assemble six faces and twelve joints exactly like the previous model.

A face complete with four joint units and an optional insert.

Two views of Windmill Base Cube 2 with inserts.

Whipped Cream Star

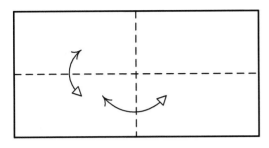

1. Start with 1:2 paper and fold and unfold both bookfolds.

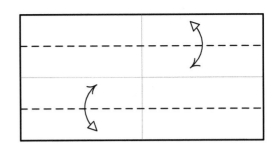

2. Fold and unfold as shown.

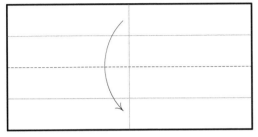

3. Re-crease existing fold.

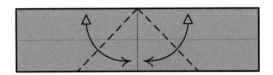

4. Fold and unfold as shown.

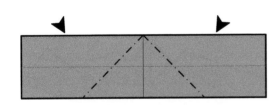

5. Inside reverse fold along creases formed in previous step.

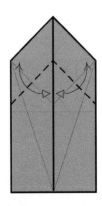

6. Mountain and valley fold as shown moving point A to the top.

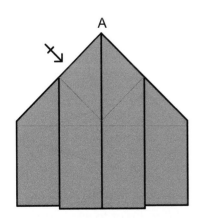

7. Repeat Step 6 at the back.

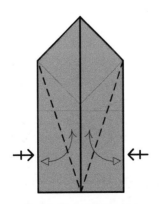

8. Valley fold and unfold top flaps. Repeat behind.

9. Valley fold and unfold as shown.

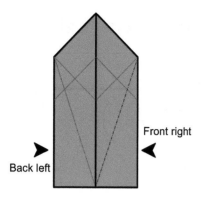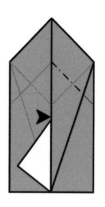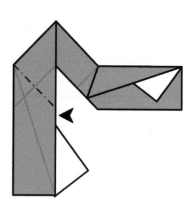

10. Inside reverse fold front right and back left flaps.

11. Inside reverse fold along crease made in Step 9.

12. Similarly, inside reverse fold left side.

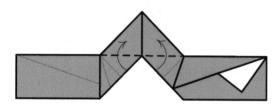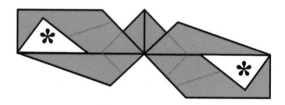

13. Valley fold top layers of both sides up.

14. Tuck white flaps in layer underneath.

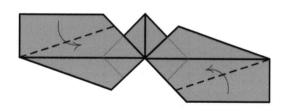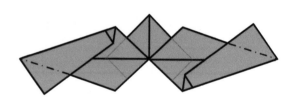

15. Valley fold along existing creases through all layers.

16. Mountain fold and unfold the extras along edges and then turn over.

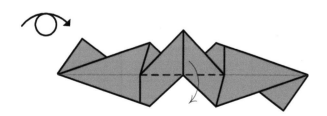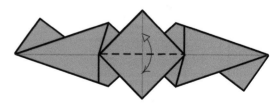

17. Valley fold downwards only the first of the three flaps.

18. Valley fold next flap downwards and orient it to point toward you.

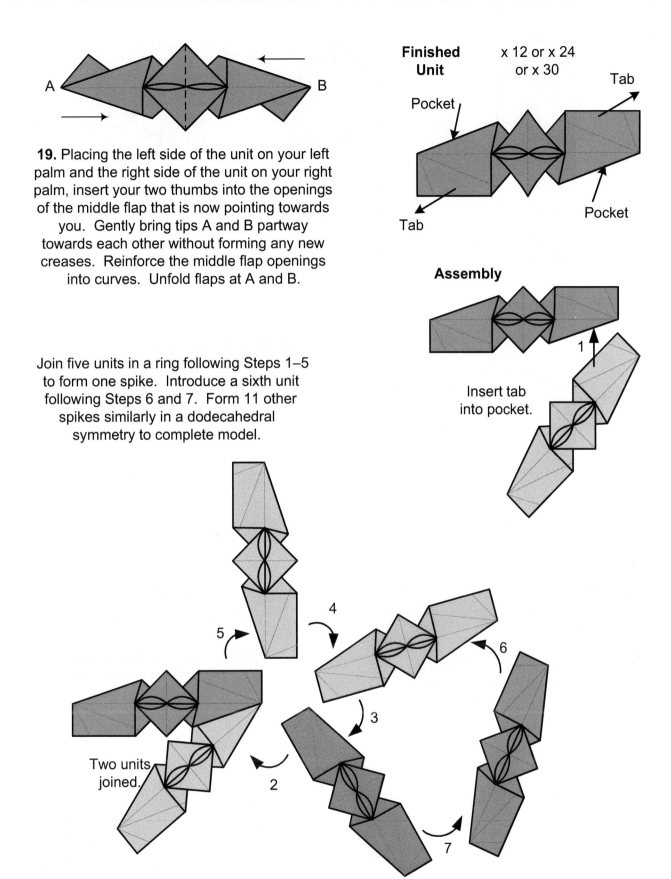

19. Placing the left side of the unit on your left palm and the right side of the unit on your right palm, insert your two thumbs into the openings of the middle flap that is now pointing towards you. Gently bring tips A and B partway towards each other without forming any new creases. Reinforce the middle flap openings into curves. Unfold flaps at A and B.

Finished Unit x 12 or x 24 or x 30

Pocket

Tab

Tab

Pocket

Assembly

Insert tab into pocket.

1

Join five units in a ring following Steps 1–5 to form one spike. Introduce a sixth unit following Steps 6 and 7. Form 11 other spikes similarly in a dodecahedral symmetry to complete model.

Two units joined.

5   4   6   3   2   7

The above shows a lesser stellated dodecahedron-like assembly. It is possible to do the greater stellated dodecahedron-like assembly (see page ii), but one must be aware that there is a lot of tension in the model and the spikes may be difficult to orient symmetrically.

Whipped Cream Star (top) and possible variations by Tanya Vysochina (bottom). Variations are not diagrammed but readers are encouraged to try for themselves. Also see page ii for another 30-unit assembly of the star.

Star with Spirals

Do Steps 1–18 of the previous model, the Whipped Cream Star (see page 69).

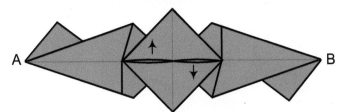

19. With your nails, curve the flap that is pointing towards you like a sine wave with left side going upwards and right side going downwards.

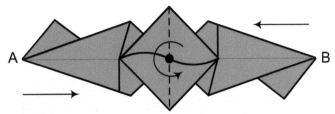

20. Holding the tip at the center of the unit that is pointing towards you (marked with a dot), firmly twist clockwise. As you do this, A and B will come partway towards each other. Unfold flaps at A and B.

Finished Unit

x 12 or x 24 or x 30

Assemble like Whipped Cream Star. Both stellations are possible for the 30-unit assembly.

Pocket

Tab

Tab

Pocket

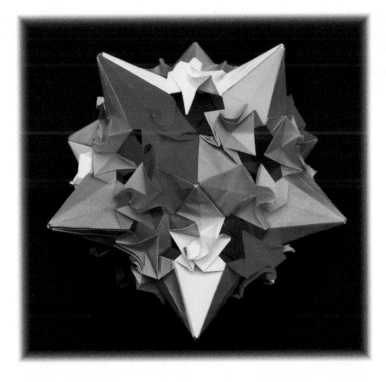

Star with Spirals in lesser stellated dodecahedron-like assembly.

Star with Spirals in greater stellated dodecahedron-like assembly (top) and a close-up view (bottom).

Star with Spirals

Whipped Cream Polyhedra

Start with 1:2 paper and do Steps 1–7 of the Whipped Cream Star on page 69.

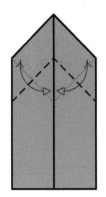

8. Valley fold and unfold as shown.

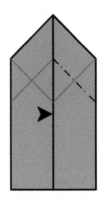

9. Inside reverse fold along crease made in Step 8.

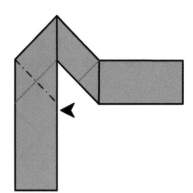

10. Similarly, inside reverse fold left side.

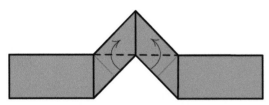

11. Valley fold top layers of both sides up.

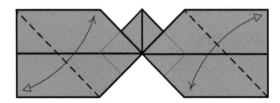

12. Valley fold and unfold.

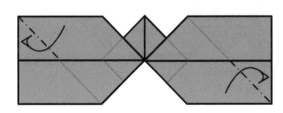

13. Mountain fold the corners in.

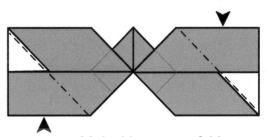

14. Inside reverse fold.

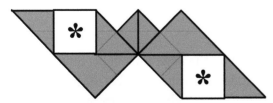

15. Tuck white flaps in layer underneath.

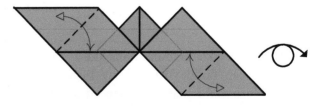

16. Fold and unfold tabs. Turn over.

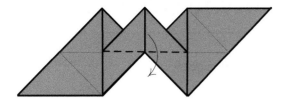

17. Valley fold downwards only the first of the three flaps.

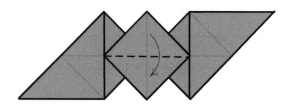

18. Swivel down next flap so that it points towards you.

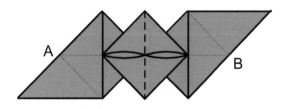

19. Repeat Step 19 of Whipped Cream Star.

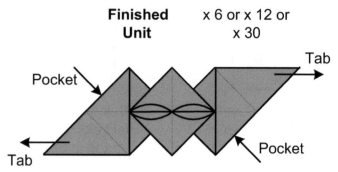

Finished Unit   x 6 or x 12 or x 30

Assembly

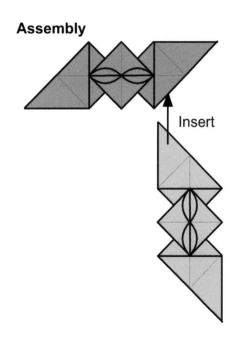

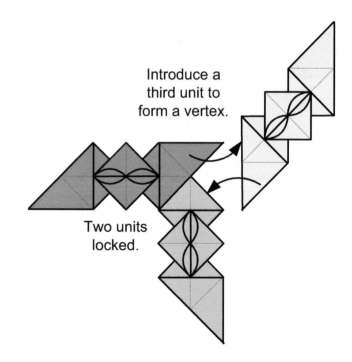

This unit is an edge unit and joins like a Sonobe unit. One can build virtually any Platonic or Archimedean solid that has three edges meeting at a vertex, e.g., tetrahedron, cube, truncated octahedron, and dodecahedron.

To make a cube, assemble 12 units forming vertices as explained in the diagram above so that the units lie on the edges of a cube. Similarly, to make a dodecahedron, assemble 30 units along the edges of a dodecahedron forming vertices as above.

Note that you may also try the spiral center version as shown in the previous model.

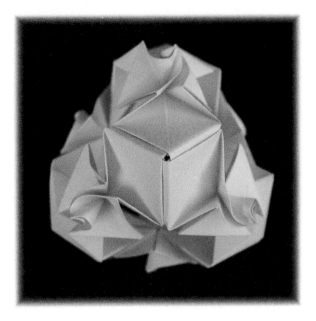

Whipped Cream Cube (left) and Tetrahedron with spiral center units (right).

Whipped Cream Dodecahedron.

Waves

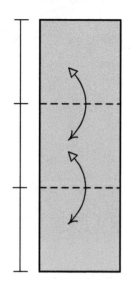

1. Start with 1:3 paper and valley fold and unfold into thirds as shown.

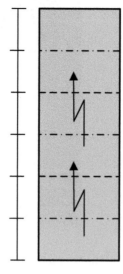

2. Pleat fold into sixths i.e., mountain and valley fold alternately.

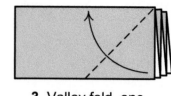

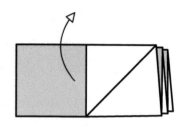

3. Valley fold, one single layer only.

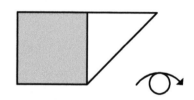

4. Unfold top flap as shown.

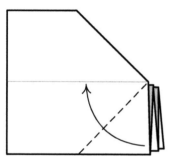

5. Valley fold all layers.

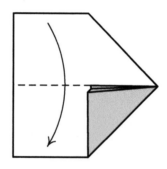

6. Re-crease valley fold shown.

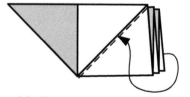

7. Turn over laterally.

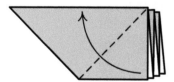

8. Valley fold all layers.

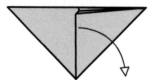

9. Unfold all layers but last.

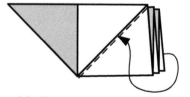

10. Re-crease valley fold on all layers and tuck underneath.

thumb   index finger

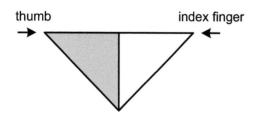

11. Apply pressure to the corners with your thumb and index finger to open up the top. Firmly curve the top edges outwards with your nails. The center flap will automatically take the shape of a sine wave. Gently reinforce it.

Finished Unit

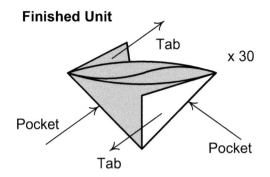

Tab

x 30

Pocket

Tab

Pocket

Assembly

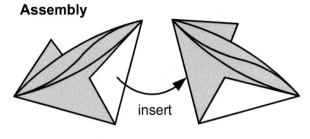

insert

Insert tab into pocket.

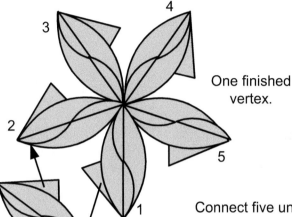

3

4

2

One finished
vertex.

5

6

1

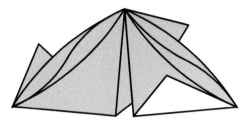

Two units joined. Make sure the top
end of the tab is all the way in. The
bottom should remain a little out.

Connect five units in a ring to complete one vertex of an
icosahedron. Introduce a sixth unit to complete one face.
Continue assembling the units along the edges of an
icosahedron to finish your model. Note: Assembly aids
such as miniature clothespins would be helpful.

Waves.

Three Interlocked Triangular Prisms (top) and Camellia (bottom).

6 ◆ A Collection of Models by Guests

In this chapter I present work by four prolific modular origami designers. Their work may already be familiar to some of you, either through the Internet or through various origami society publications. I am introducing them in this book with the hope that their wonderful work will be exposed to the much wider audience that it truly deserves. I thank my guest creators immensely for generously contributing their designs to this book.

Paper type and size recommendations for the models in this chapter are not being made here, though some of the individual models include such recommendations in their instructions. Also, in order to preserve the guest contributors' diagramming styles, the symbols used by them have not been changed. There may be a different style of numbering the steps. In certain cases valley folds have been shown in red. Some of the alternate symbols used by the guests are shown below.

	Symbol used earlier	Alternate Symbol
Valley fold	- - - - - - - - -	– – – – – – – –
Mountain fold	–·–·–·–·–	···–···–···–
Fold and unfold	⤹	⤸

Lily of the Nile model embellished with beads.

Truncated Rhombic Triacontahedron

Daniel Kwan

This model requires 30 identical units.
Using 3" squares results in a ball approximately 5" in diameter.

1.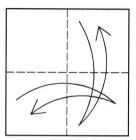

Start white side up. Fold and unfold in half in both directions.

2.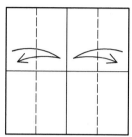

Fold and unfold two opposite edges to the center.

3.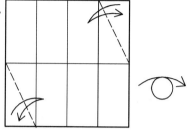

Fold and unfold the top right and lower left corners as shown. Turn over.

4.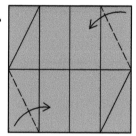

Fold in the other two corners as shown.

5.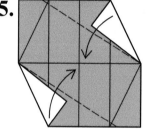

Pivoting around the endpoints of the horizontal half fold, bring the folded edges to the center crease.

6.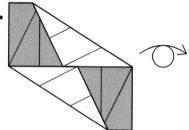

It should look like this. Turn over.

7.

Refold along the two creases from Step three.

8.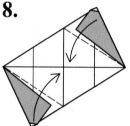

Pivoting around the same corners in Step 5, bring the folded edges from Step 7 to the center crease.

9.

Inside reverse fold the corners to the center using the existing crease below.

10.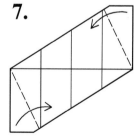

Unfold back to Step 7 and turn over.

11.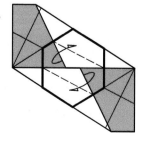

Mountain fold and unfold along the triangular flaps as shown so that the ends of these creases lie on the border of the bold hexagon. This additional fold will make the assembly locking process easier. Fold 29 more identical units.

82

Daniel Kwan

12.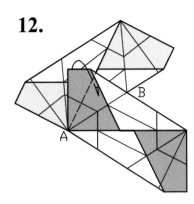

Locking together two units is a two step process. First, overlay one over the other as shown on the left. Note how the creases line up at points A and B. Wrap the end flap from the top unit around the triangular flap of the other unit using the crease from Step 7.

13.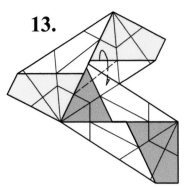

The second step is to fold back along the triangular flap through all layers along the fold we made in Step 11.

14.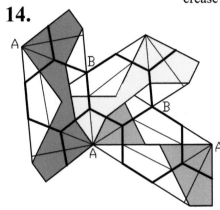

As you assemble more units, make sure you have five units joined around A-corners and three around B-corners. As you complete A-corners, make sure the corner is pushed down to form a dimpled pentagon along the existing creases. In this image, the bolded lines represent the creases that will become the edges of the completed shape.

15.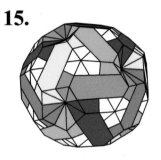

Completed Truncated Rhombic Triacontahedron.

Truncated Rhombic Triacontahedron (see also page ii).

Four Interlocked Triangular Prisms

Daniel Kwan

6" paper makes a 5" diameter model.
Start with four colors, using three square sheets of each color (twelve sheets total).

Paper Preparation

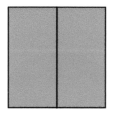 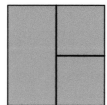 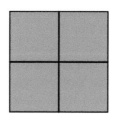

For each color, cut or tear the three squares into three 2×1's and six 1×1's. None of the torn edges will show on the completed model. The 2×1's are for Unit A and the 1×1's are for Unit B.

Unit A

1.

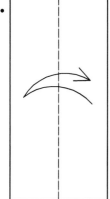

Start white side up. Fold and unfold in half.

2.

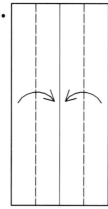

Fold both sides to the center.

3.

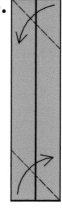

Fold and unfold opposite corners.

4.

Inside reverse fold both corners underneath the side flaps. Leave the small triangles unfolded.

5.

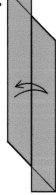

Fold and unfold the unit in half lengthwise.

6.

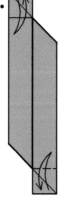

Fold and unfold the ends to form square-shaped tabs. Turn over.

7.

Completed Unit A. Fold the other 11.

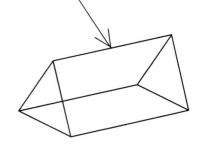

Unit A's will become the three parallel long edges of each triangular prism.

1.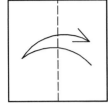

Start white side up.
Fold and unfold in half.

2.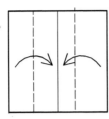

Fold both sides
to the center.

3.

Fold and unfold the
lower left corner to
the center.

4.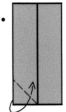

Inside reverse
fold on the
previous crease.

5.

Bring the right
side flap out to
the right edge
and make a
pinch near the
bottom.

6.

Pivot around the
center of the
bottom to bring
the lower right
corner to lie on
the pinch mark.

7.

It should look
like this.
Unfold and
rotate 180°.

8.

Pivot around the
lower left corner to
bring the lower right
corner to the center of
the unit. Pinch only
on the right edge.

9.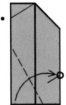

Fold the lower
left corner to the
right so the point
lies where the
pinchmark hits
the right edge.

10.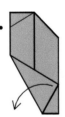

This should
be the result.
Unfold the
last fold.

11.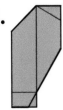

Inside reverse fold
the lower left corner
underneath the right
side flap. Keep the
tiny corner on the
bottom unfolded.

12.

Fold and
unfold the unit
in half down
the center.

13.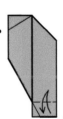

Fold and unfold
the bottom end
to form a
rectangular tab.
Turn over.

14.

Completed Unit B.
Fold the other 23.

Unit B's will become
the six triangle-forming
short edges of each
triangular prism.

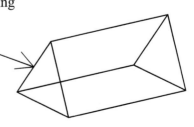

1.

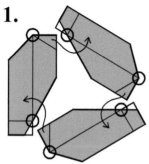

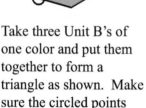

Take three Unit B's of one color and put them together to form a triangle as shown. Make sure the circled points end up together at every connection.

2.

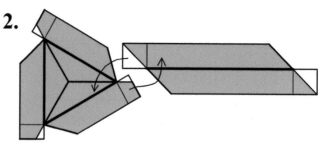

Add each of the three Unit A's for this color to the corners of the triangle, one at a time. When adding each Unit A, it is important that the tab goes underneath the side flap on the far side of the Unit B's pocket in addition to the reverse fold pocket (it will not always go there automatically). The model should become three dimensional after adding the first Unit A. In this diagram, the darkened edges represent the edges of the completed prism. Make sure the dark edge of the Unit A is perpendicular to both adjoining dark edges of the Unit B's.

3.

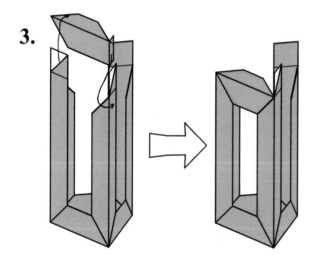

To complete the second triangular end-cap to the prism, start by adding each of the last three Unit B's one at a time. Each Unit B bridges a connection between two Unit A's. Just like in Step 2, make sure the Unit A tab goes underneath the side flap on the far side of the pocket.

5.

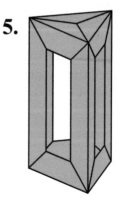

4.

Once the last three Unit B's are in place, lock them into each other all at once to complete the prism.

This is one completed triangular prism. Repeat assembly Steps 1 and 2 for the other prisms, leaving the second end-caps off.

6.

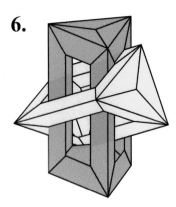

Add each additional prism by weaving it through the already assembled one(s) and then attaching the three Unit B's to complete the second triangular end-cap. Two assembled prisms should look like this. Every pair combination of prisms is interlocked in this way with two long bars from each prism threading through the other.

7.

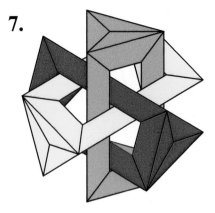

This is three of the four prisms. When you thread the third prism through, look for the two three-bar woven intersections on opposite sides of the model, as shown here in the centre of this diagram. After adding the third prism, the model should be sturdy.

8.

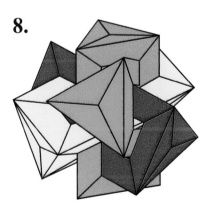

To add the fourth prism, just thread the three long struts through the three holes seen in the previous diagram. This is the completed model.

Four Interlocked Triangular Prisms (see also page 80).

Chrysanthemum Leroy

Carlos Cabrino

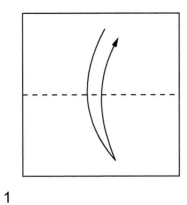

1

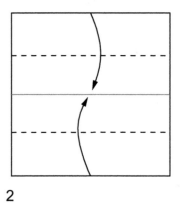

2

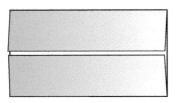

3

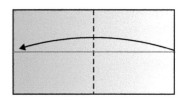

4

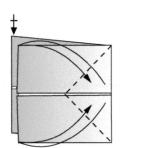

5   Repeat on the other side.

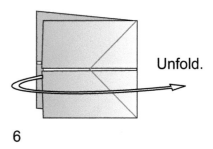

Unfold.

6

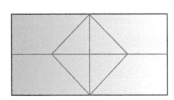

7

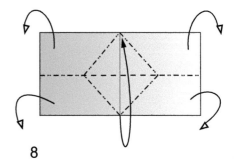

8

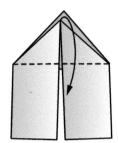

9

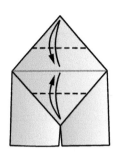

10

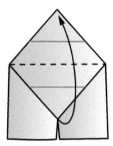

11

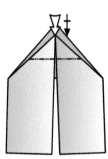

12   Inside reverse fold and
      repeat on the other side.

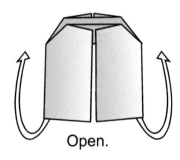

Open.

13

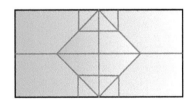

14

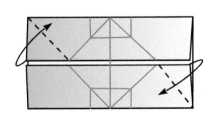

15

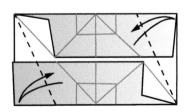

16

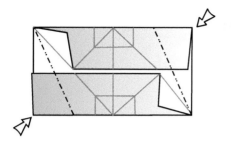

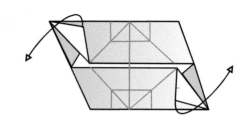

17 Inside reverse fold corners.

18

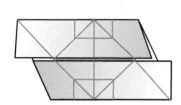

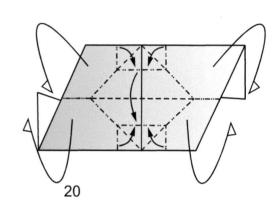

19

20

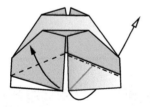

21

22

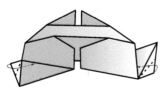

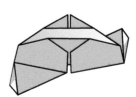

23

24

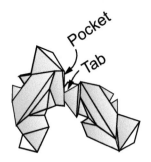

12-unit cube assembly.

30-unit dodecahedral assembly.

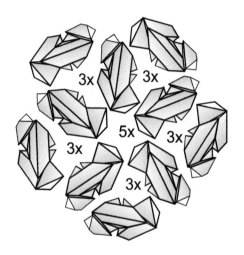

30- and 12-unit assemblies of Chrysanthemum.

Chrysanthemum Leroy Variation

Carlos Cabrino

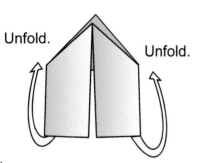

Unfold.   Unfold.

1   Repeat Steps 1–9 of the Chrysanthemum Leroy.

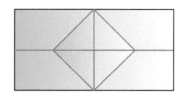

2

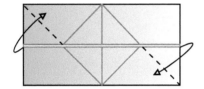

3

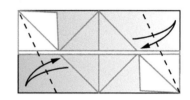

4

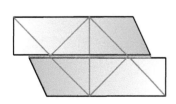

5   Inside reverse fold corners.

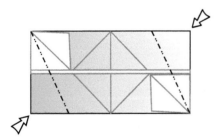

6

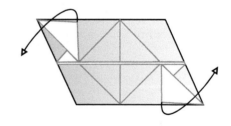

7

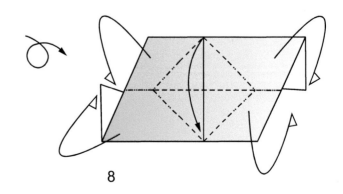

8

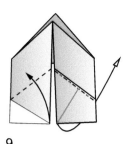

9

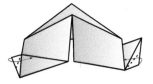

10

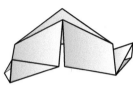

Tab

Pocket

11

12-unit cube assembly.

30-unit dodecahedral assembly.

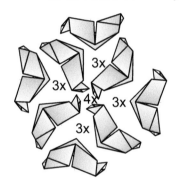

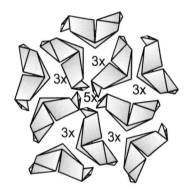

Chrysanthemum variation.

Carnation Leroy

Carlos Cabrino

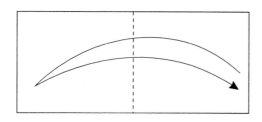

1 Start with 1:2 paper.

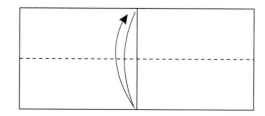

2

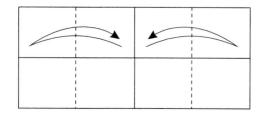

3

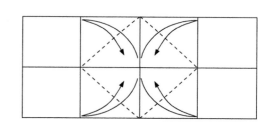

4

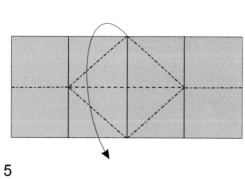

5

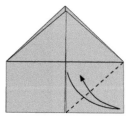

6

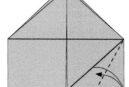

7

8

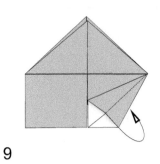

9

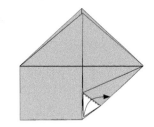

10

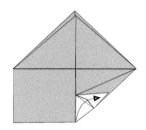

11

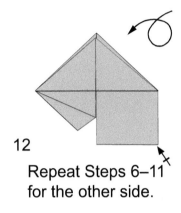

12

Repeat Steps 6–11
for the other side.

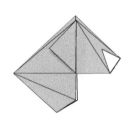

13

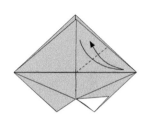

14

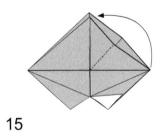

15

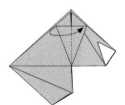

16

17

18

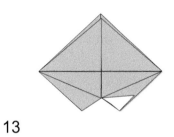

19

20

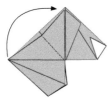

21 22 23

24

25 Repeat Steps 13–24
on the reverse side.

26

30-unit dodecahedral assembly.

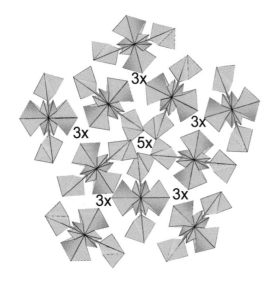

Carnation

Camellia

Tanya Vysochina

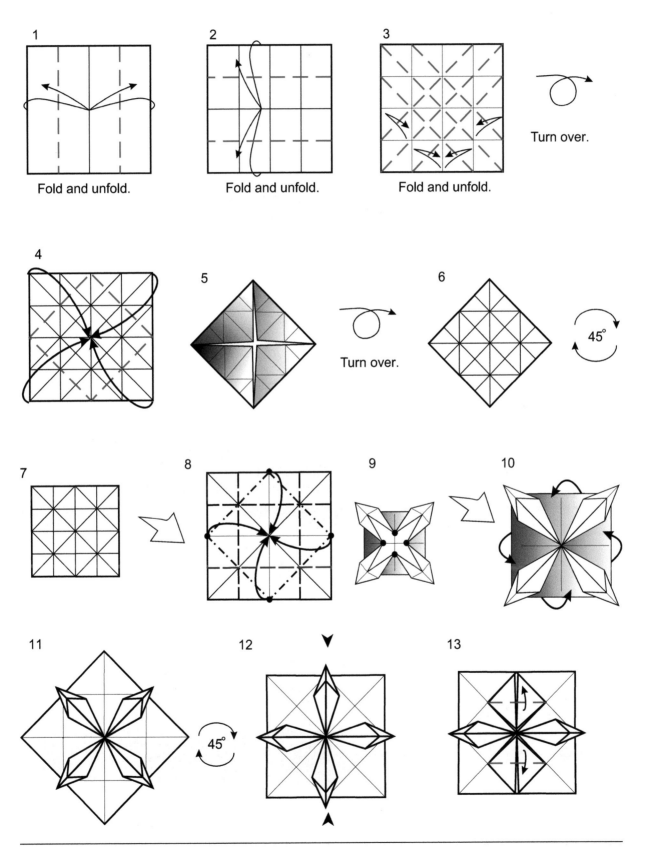

1
Fold and unfold.

2
Fold and unfold.

3
Fold and unfold.

Turn over.

4

5

Turn over.

6

45°

7

8

9

10

11

45°

12

13

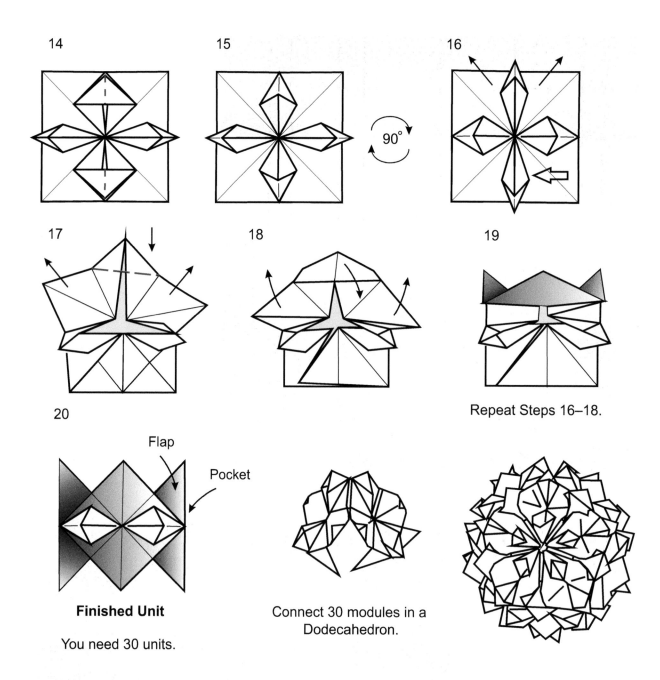

14 **15** **16**

90°

17 **18** **19**

Repeat Steps 16–18.

20

Flap

Pocket

Finished Unit

You need 30 units.

Connect 30 modules in a
Dodecahedron.

Variation

Begin with Step 20 above.

20

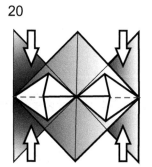

21

Finished Unit

You need 30 units.

Connect 30 modules in a
Dodecahedron.

Tanya Vysochina

Camellia (top) and Camellia variation (bottom). (See also page 80.)

Dahlia

Unit A

1

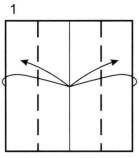

Fold and unfold.

2

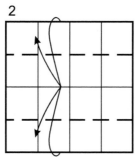

Fold and unfold.

3

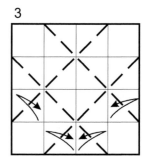

Fold and unfold.

4

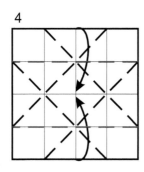

5

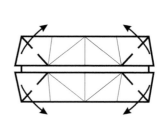

6

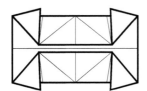

Turn over.

7

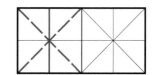

8

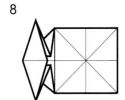

9

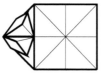

Repeat Steps 7–9.

10

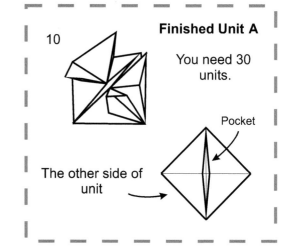

Finished Unit A

You need 30 units.

The other side of unit

Pocket

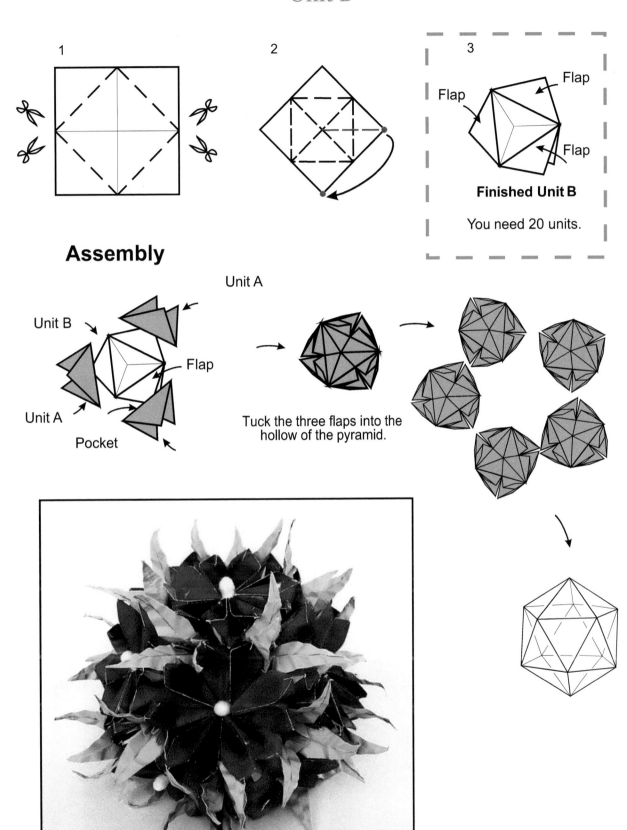

1

2

3

Flap

Flap

Flap

Finished Unit B

You need 20 units.

Assembly

Unit A

Unit B

Flap

Unit A

Pocket

Tuck the three flaps into the
hollow of the pyramid.

Dahlia with leaf inserts.
(Leaf inserts not diagrammed.)

Begin with Step 8 on page 100.

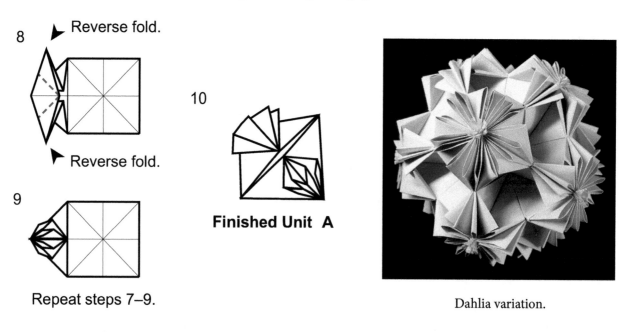

8 ▶ Reverse fold.

▶ Reverse fold.

10

9

Finished Unit A

Repeat steps 7–9.

Dahlia variation.

Four other possible petal finishes for the Dahlia Unit A. The variations are minor and hence not diagrammed. Readers are encouraged to try these variations.

Tanya Vysochina

Lily of the Nile

Tanya Vysochina

Unit A

Begin with Step 8 on page 100.

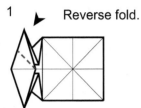

1 Reverse fold.

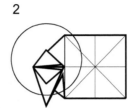

2

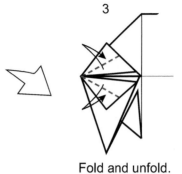

3

Fold and unfold.

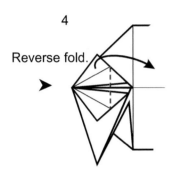

4 Reverse fold.

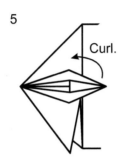

5 Curl.

Repeat Steps 2–5 with other three parts.

6 **Finished Unit A**

You need 30 units.

Pocket

The other side of unit

Unit B

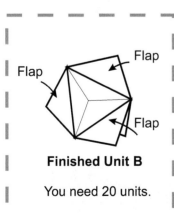

Flap

Flap

Flap

Finished Unit B

You need 20 units.

Assembly

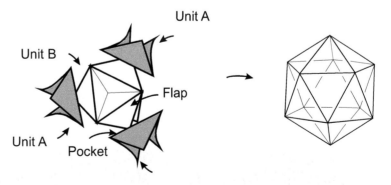

Unit A

Unit B

Flap

Unit A

Pocket

Connect 20 triangular modules in an icosahedron.

See page 81 for finished model photo.

Tanya Vysochina

1

Fold and unfold.

2

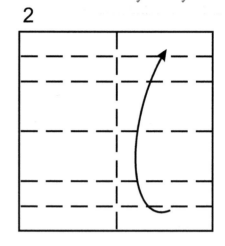

3

4

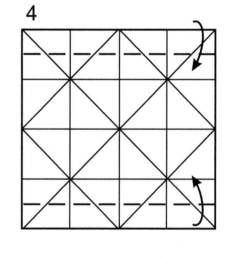

5

6

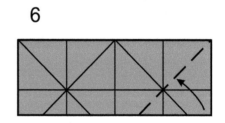

7

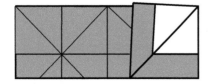

Turn over.

8

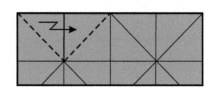

9

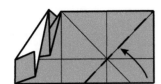

10

11

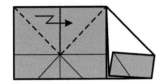

12

13

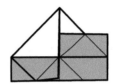

14

15

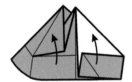

16

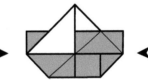

Turn over.

17

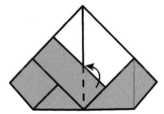

18

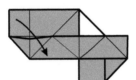

19

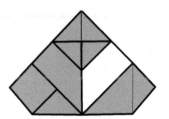

20

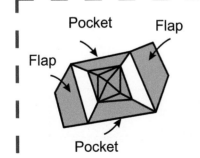

21

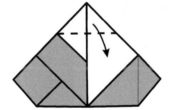

22

Repeat Steps 20–21
on the other side.

Pocket   Flap

Flap

Finished
Module

Pocket

Assembly

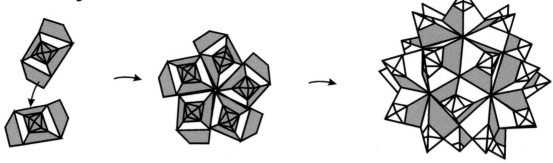

Connect 30 modules in a dodecahedron.

Variation A

Begin with Step 22 on page 105.

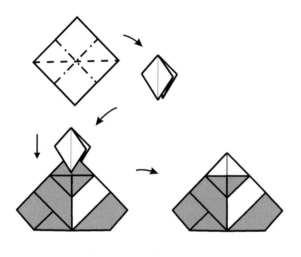

Variation B

Begin with Step 22 on page 105.

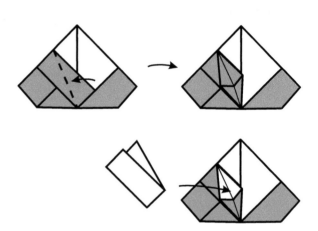

Crystal Variation A (left) and Crystal Variation B without inserts (right).

Adaptable Dodecahedron

Aldo Marcell

(Diagrams by Meenakshi Mukerji)

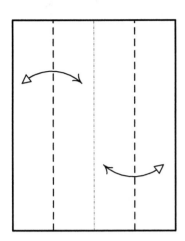

1. Start with 4:5 paper. Fold and unfold all the creases shown.

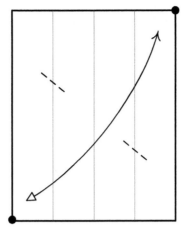

2. Bring dot to dot and pinch the two points shown.

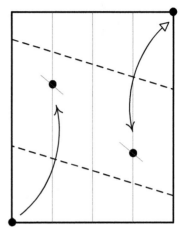

3. Bring dot to dot; fold and unfold top crease and fold bottom crease.

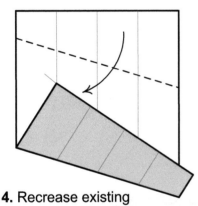

4. Recrease existing crease and turn over.

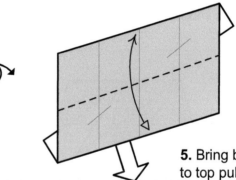

5. Bring bottom edge to top pulling flap out. Unfold all.

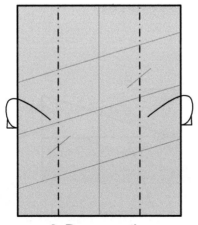

6. Recrease the mountain folds shown.

7. Valley fold along existing crease.

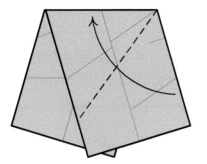

8. Valley fold to bring edge to edge.

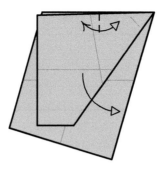

9. Pinch to trace the center of the back edge. Unfold Step 8.

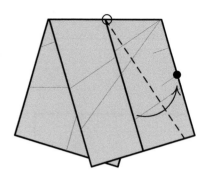

10. Valley fold to bring flap edge to dot.

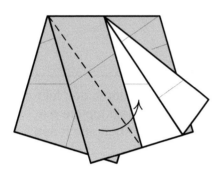

11. Valley fold as shown.

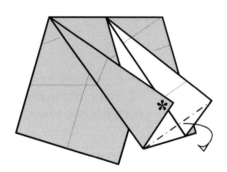

12. Tuck flap underneath and mountain fold tip.

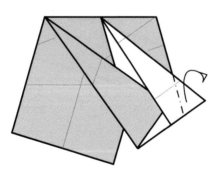

13. Mountain fold along edge and unfold.

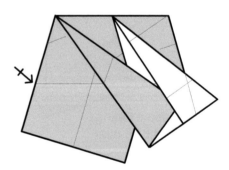

14. Repeat Steps 8–13 on the reverse.

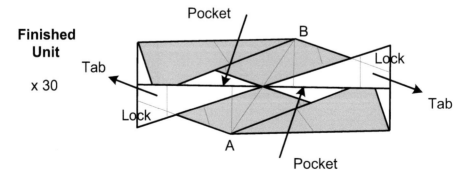

Finished Unit

x 30

Pocket

B

Lock

Tab

Tab

Lock

A

Pocket

Note: Line AB will lie along the edge of the finished dodecahedron.

(Diagrams by Meenakshi Mukerji)

Assembly

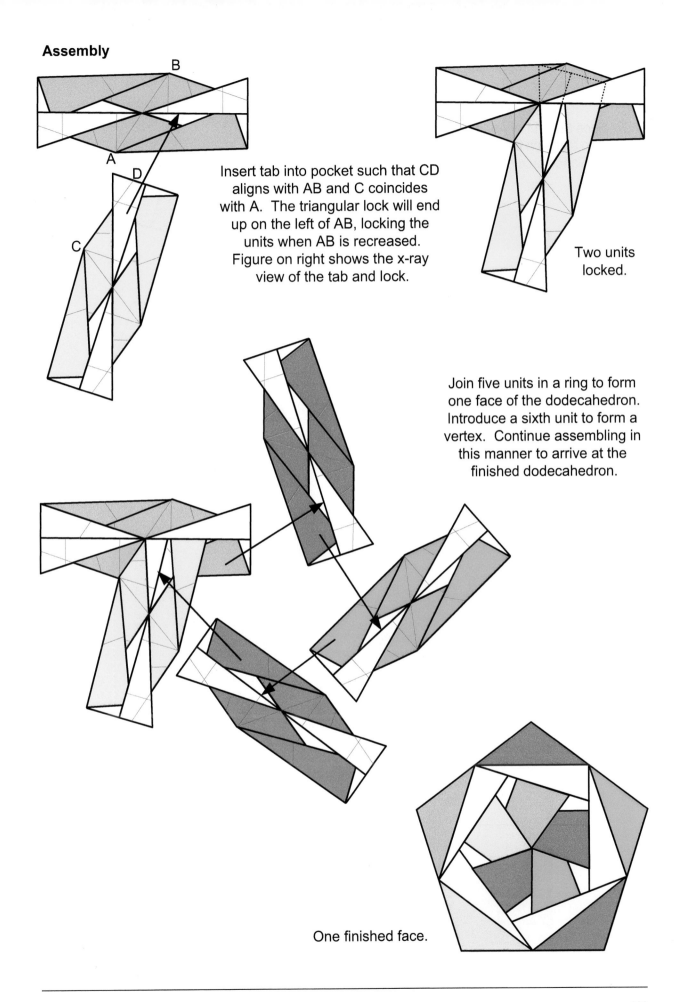

Insert tab into pocket such that CD aligns with AB and C coincides with A. The triangular lock will end up on the left of AB, locking the units when AB is recreased. Figure on right shows the x-ray view of the tab and lock.

Two units locked.

Join five units in a ring to form one face of the dodecahedron. Introduce a sixth unit to form a vertex. Continue assembling in this manner to arrive at the finished dodecahedron.

One finished face.

Variation

Replace Step 11 of Adaptable Dodecahedron with
the two steps on the right to narrow the left strip.
One finished face of the variation is shown below.

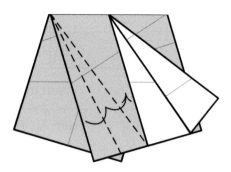

11a.

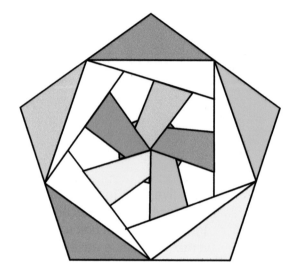

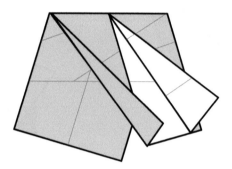

11b.

Adaptable Dodecahedron.

Aldo Marcell

Aldo Marcell
(Diagrams by Meenakshi Mukerji)

Start by completing up to Step 11 of
Adaptable Dodecahedron.

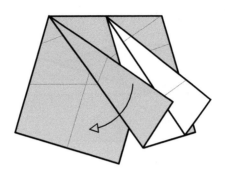

12. Unfold flap.

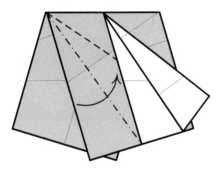

13. Squash fold.

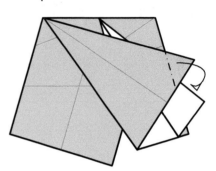

14. Mountain fold along edge.

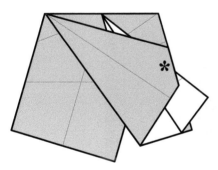

15. Tuck flap underneath.

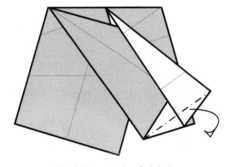

16. Mountain fold the
end of the tab.

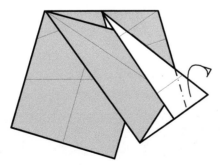

17. Mountain fold and
unfold to form lock.

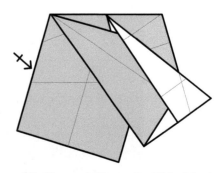

18. Repeat Steps 8–17 behind.

**Finished
Unit**

x 30

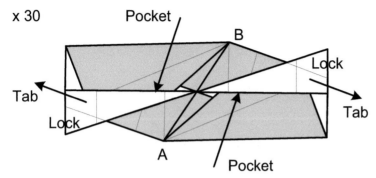

Pocket

B

Lock

Tab

Lock

Tab

A

Pocket

Assemble exactly like Adaptable
Dodecahedron. One finished
face shown below.

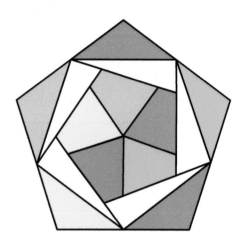

Adaptable Dodecahedron 2.

Appendix

Bibliography and Suggested Reading

◈ [Bee01] Rick Beech, *Origami: The Complete Practical Guide to the Ancient Art of Paperfolding*, Lorenz Books, 2001.

◈ [Bos03] *British Origami Society, BOS Magazine 219, April 2003*.

◈ [Cox73] H. S. M. Coxeter, *Regular Polytopes*, Reprinted by Dover Publications, 1973.

◈ [Dem07] Erik D. Demaine & Joseph O' Rourke, *Geometric Folding Algorithms: Linkages, Origami, Polyhedra*, Cambridge University Press, 2007.

◈ [Dir97] Alexandra Dirk, *Origami Boxes for Gifts, Treasures and Trifles*, Sterling, 1997.

◈ [Fer07] Bruno Ferraz, *Ultrapassando Fronteiras com o Origami (Exceeding Borders with Origami)* (in Portugese), Editora Ciência Moderna, 2007.

◈ [Fus89] Tomoko Fuse, *Origami Boxes*, Japan Publications Trading, 1989.

◈ [Fus90] Tomoko Fuse, *Unit Origami: Multidimensional Transformations*, Japan Publications, 1990.

◈ [Fus92] Tomoko Fuse, *Let's Fold Spirals*, Chikuma Shobo, 1992.

◈ [Fus95] Tomoko Fuse, *Origami Spirals*, Chikuma Shobo, 1995.

◈ [Fus96] Tomoko Fuse, *Joyful Origami Boxes*, Japan Publications Trading, 1996.

◈ [Fus98] Tomoko Fuse, *Fabulous Origami Boxes*, Japan Publications Trading, 1998.

◈ [Fus00] Tomoko Fuse, *Quick and Easy Origami Boxes*, Japan Publications Trading, 2000.

◈ [Fus02] Tomoko Fuse, *Kusudama Origami*, Japan Publications Trading, 2002.

◈ [Fus06] Tomoko Fuse, *Unit Polyhedron Origami*, Japan Publications Trading, 2006.

◈ [Fus07] Tomoko Fuse, *Floral Origami Globes*, Japan Publications Trading, 2007.

◈ [Gil07] Eduardo Gil Moré, *Papiroflexia y Geometría* (in Spanish), Miguel A Salvatella, 2007.

◈ [Gje08] Eric Gjerde, *Origami Tessellations: Awe-Inspiring Geometric Designs*, A K Peters, Ltd., 2008.

◈ [Gra76] Alice Gray, "On Modular Origami," *The Origamian* vol. 13, no. 3, page 4, June 1976.

◈ [Gro05] Gay Merrill Gross, *Origami: An Origami Christmas at Your Fingertips*, Barnes & Noble, 2005.

◈ [Gur95] Rona Gurkewitz and Bennett Arnstein, *3-D Geometric Origami: Modular Polyhedra*, Dover Publications, 1995.

◈ [Gur99] Rona Gurkewitz, Bennett Arnstein, and Lewis Simon, *Modular Origami Polyhedra*, Dover Publications, 1999.

◈ [Gur03] Rona Gurkewitz and Bennett Arnstein, *Multimodular Origami Polyhedra*, Dover Publications, 2003.

◈ [Gur08] Rona Gurkewitz, *Beginner's Book of Multimodular Origami Polyhedra: The Platonic Solids*, Dover Publications, 2008.

◈ [Hul02] Thomas Hull, ed., *Origami 3: Third International Meeting of Origami Science, Mathematics, and Education*, A K Peters, Ltd., 2002.

◈ [Hul06] Thomas Hull, *Project Origami: Activities for Exploring Mathematics*, A K Peters, Ltd., 2006.

◈ [Jac87] Paul Jackson, *Encyclopedia of Origami/Papercraft Techniques*, Headline, 1987.

◈ [Jac89] Paul Jackson, *Origami: A Complete Step-by-step Guide*, Hamlyn, 1989.

◈ [Kas98] Kunihiko Kasahara, *Origami for the Connoisseur*, Japan Publications, 1998.

◈ [Kasa98] Kunihiko Kasahara, *Origami Omnibus: Paper Folding for Everybody*, Japan Publications, 1998.

◈ [Kas03] Kunihiko Kasahara, *Extreme Origami*, Sterling, 2003.

◈ [Kaw70] Toyoaki Kawai,*Origami*, Nursing, Inc., New Edition, 1970.

◈ [Kaw02] Miyuki Kawamura, *Polyhedron Origami for Beginners*, Japan Publications, 2002.

◈ [Kaw01] Toshikazu Kawasaki, *Origami Dream World* (in Japanese), Asahipress, 2001.

◈ [Kaw05] Toshikazu Kawasaki, *Roses, Origami & Math*, Japan Publications Trading, 2005.

◈ [Kla09] Robert Klanten, *Papercraft: Design and Art With Paper*, Die Gestalten Verlag, 2009.

◈ [Lan03] Robert Lang, *Origami Design Secrets: Mathematical Methods for an Ancient Art*, A K Peters, 2003.

◈ [Lan08] Robert Lang, ed., *Origami 4: Fourth International Meeting of Origami Science, Mathematics, and Education*, A K Peters, Ltd., 2009.

◈ [Mit97] David Mitchell, *Mathematical Origami: Geometrical Shapes by Paper Folding*, Tarquin, 1997.

◈ [Mit00] David Mitchell, *Paper Crystals: How to Make Enchanting Ornaments*, Water Trade, 2000.

◈ [Mon09] John Montroll, *Origami Polyhedra Design*, A K Peters, Ltd., 2009.

◈ [Muk07] Meenakshi Mukerji, *Marvelous Modular Origami*, A K Peters, Ltd., 2007.

◈ [Muk08] Meenakshi Mukerji, *Ornamental Origami: Exploring 3D Geometric Designs*, A K Peters, Ltd., 2008.

◈ [NOA94] NOA, *Minna Kusudama*, Nihon Origami Kyokai, 1994.

◈ [Nol95] J. C. Nolan, *Creating Origami*, Alexander Blace & Co., 1995.

◈ [Ori00] Origami Sociëteit Nederland, *Orison Magazine 16/03*, May 2000.

◈ [Ow96] Francis Ow, *Origami Hearts*, Japan Publications, 1996.

◈ [Pet98] David Petty, *Origami Wreaths and Rings*, Aitoh, 1998.

◈ [Pet02] David Petty, *Origami 1-2-3*, Sterling, 2002.

◈ [Pet06] David Petty, *Origami A-B-C*, Sterling, 2006.

◈ [Rob04] Nick Robinson, *The Encyclopedia Of Origami*, Running Press, 2004.

◈ [Row66] Tandalam Sundara Row, *Geometric Exercises in Paper Folding*, Reprinted by Dover Publications, 1966.

◈ [Tan02] Origami Tanteidan,*Origami Tanteidan Convention No.8*, Origami House, 2002.

◈ [Tem86] Florence Temko, *Paper Pandas and Jumping Frogs*, China Books & Periodicals, 1986.

◈ [Tem04] Florence Temko, *Origami Boxes and More*, Tuttle Publishing, 2004.

◈ [Tub 06] Arnold Tubis and Crystal Mills, *Unfolding Mathematics with Origami Boxes*, Key Curriculum Press, 2006.

◈ [Tub07] Arnold Tubis and Crystal Mills, *Fun with Folded Fabric Boxes*, C&T Publishing, 2007.

◈ [Yam90] Makoto Yamaguchi, *Kusudama Ball Origami*, Japan Publications, 1990.

Suggested Websites

◈ [Ada] Sara Adams, *Happy Folding*, http://www.happyfolding.com/

◈ [Aha] Gilad Aharoni, *Gilad's Origami Page*, http://www.giladorigami.com/

◈ [And] Eric Andersen, *paperfolding.com*, http://www.paperfolding.com/

◈ [Bos] British Origami Society, *BOS Home Page*, http://britishorigami.info/

◈ [Bur] Krystyna Burczyk, *Krystyna Burczyk's Orig-ami Page*, http://www1.zetosa.com.pl/~burczyk/origami/index-en.html (English version)

◈ [Cab] Carlos Cabrino, *Leroy—Origami*, http://origamileroy.spaces.live.com/

◈ [Har] George W. Hart, *The Pavilion of Polyhedreality*, http://www.georgehart.com/pavilion.html

◈ [Hul] Tom Hull, *Tom Hull's Home Page*, http://mars.wnec.edu/~th297133/

◈ [Jap] Japan Origami Academic Society, *Origami Tanteidan*, http://www.origami.gr.jp/index-e.html

◈ [Kat] Rachel Katz, *Origami with Rachel Katz*, http://www.origamiwithrachelkatz.com

◈ [Kwa] Daniel Kwan, *Daniel Kwan's Photostream*, http://www.flickr.com/photos/8303956@N08/

◈ [Lan] Robert J. Lang, *Robert J. Lang Origami*, http://www.langorigami.com/

◈ [Luk] Ekaterina Lukasheva, *Kusudama Me!*, http://www.kusadama.me/

◈ [Mar] Aldo Marcell, *Aldo Marcell*, http://origami.artists.free.fr/AldoMarcell/

◈ [Mit] David Mitchell, *Origami Heaven*, http://www.origamiheaven.com/

◈ [Muk] Meenakshi Mukerji, *Origami—MM's Modular Mania*, http://www.origamee.net/

◈ [Ori] Origami Resource Center, *Origami: the Art of Paper Folding*, http://www.origami-resource-center.com/index.html

◈ [Ous] OrigamiUSA, *OrigamiUSA Home Page*, http://origami-usa.org/

◈ [Ow] Francis Ow, *Francis Ow's Origami Page*, http://web.singnet.com.sg/~owrigami

◈ [Pet] David Petty, *Dave's Origami Emporium*, http://www.davidpetty.me.uk/

◈ [Pla] Jim Plank, *Jim Plank's Origami Page (Modular)*, http://www.cs.utk.edu/~plank/plank/origami/

◈ [Rez] Jorge Rezende, *Jorge Rezende Home Page*, http://gfm.cii.fc.ul.pt/people/jrezende

◈ [Ros] Halina Rościszewska-Narloch, *Haligami*, http://www.origami.friko.pl/index3.php

◈ [Sha] Rosana Shapiro, *Modular Origami*, http://www.ulitka.net/origami/

◈ [Shu] Yuri & Katrin Shumakov, *Oriland*, http://www.oriland.com

◈ [Ter] Nicolas Terry, *Passion Origami.com*, http://www.passionorigami.com/

◈ [Ver] Helena Verrill, *Origami*, http://www.math.lsu.edu/~verrill/origami/

◈ [Vers] Paula Versnick, *Orihouse*, http://www.orihouse.com/

◈ [Vys] Tanya Vysochina, *My Kusudamas*, http://pics.livejournal.com/tigreshenka/gallery/00001xf5

◈ [Wal] Dennis Walker, *Origami Database*, http://origamidatabase.com/

◈ [Wol] Wolfram Research, *Wolfram MathWorld*, http://mathworld.wolfram.com/

◈ [Wu] Joseph Wu, *Joseph Wu Origami*, http://www.origami.vancouver.bc.ca/

About the Guest Contributors

Below are introductions to my guest contributors as written in their own words (some in first person and some in third), in no particular order.

Carlos Cabrino (Leroy)

Website: http://origamileroy.spaces.live.com/

Carlos Cabrino was born in São Caetano do Sul, São Paulo, Brazil. In 1981 he graduated in mathematics and started working in the area of Information Technology. In 1989 when he began to look at various kinds of therapy for his father who was then ill, he got interested in origami. Self-taught, he continued to fold from then on. In 2006 he began to create his own designs in modular origami. He enjoys modular origami because of the geometric and polyhedral forms that one can make and because of his love for mathematics. He shares his work on his website under the nickname Leroy.

Tanya Vysochina

Website: http://pics.livejournal.com/tigreshenka/gallery/00001xf5

Then by chance I saw on the Internet pictures of various beautiful kusudamas and very much wanted to learn to make them. I was successful and so excited that I even remember the date I made my first ever kusudama. It was December 21, 2008. I was so satisfied and happy that I began to fold more and more. Then I wanted to make an attempt to create some designs of my own rather than just follow other people's diagrams. First I started to make variations to existing models by other artists including Meenakshi Mukerji's models. You can see photos of two such variations on page 72. Then I wanted to make models that were completely my own designs. The first creation that I am proud of is my Camellia series of models and I am happy that Meenakshi has selected this for her book. I created it in August of 2009. And since then I kept designing more and more and have not stopped yet. You can see my work and find some photo diagrams on my website.

I studied in Kyiv State Economic University, now called Vadym Hetman Kyiv National Economic University, and got a master's degree in 1996. I work in area of economics.

My acquaintance with modular origami began with the Chinese modules, also called the Golden Venture Units, towards the end of the year 2008.

Daniel Kwan

Website: http://www.flickr.com/photos/8303956@N08/

I first started folding when I was five. My parents had me go to Chinese school, and one of the after-school activities there was origami. Although I dropped out of the Chinese school after only one year, I had a continuing interest in origami and I began to fold from origami books. My folding particularly jump-started when I found the book *Unit Origami* by Tomoko Fusé and discovered how much I enjoyed modular origami. My mother, being a math teacher, also started getting into origami when she saw the geometric and modular pieces I was folding. As a child, having a mother who also did origami had an immense fostering effect. She helped me greatly through both the strong support she had for the hobby and through the supplies I had gained access to. She herself began teaching origami classes to elementary and middle school students, and currently she organizes a monthly origami meeting in our hometown, and bigger biannual gatherings in our house. It is also thanks to her that I began attending the OUSA annual conventions since 1997. Seeing so many other dedicated origami artists and impressive exhibition displays had a dramatic effect on my motivation and inspiration towards origami. I didn't fully get into designing my own modular origami until 2002 when I designed the Four Triangular Prisms model as an imitation of a wooden puzzle I had. Up until June 2008, I continued designing origami modulars based on various polyhedra and polyhedra compounds. A good portion of these used edge-based units to create interwoven frames. In June 2008, I made a sudden switch of focus to origami tessellations, which I have been almost exclusively designing and folding until the present.

I started my undergraduate studies in mechanical engineering, but decided it wasn't for me and ended up graduating in 2009 with a BS in accounting from the Rutgers Business School in New Brunswick, New Jersey. My other hobbies are juggling and using other juggling related props, and playing Go, the board game. I maintain a Flickr® album to display my work and to even share some diagrams.

Aldo Marcell

Website: http://origami.artists.free.fr/AldoMarcell/

My full name is Aldo Marcel Velásquez Olivas, but I like to be recognized as Aldo Marcell. I was born in 1978 in Estelí, Nicaragua, and I live there as well. Estelí is a city in the northern part of Nicaragua. I am a biologist and a botanist and I specialize in the wild plants of Nicaragua. That's why I like to design plant representations in origami. But I also like to design modular origami. Other areas of my work include being a tourist guide in Nicaragua, participating in environmental education, and social work. Sometimes I work with the origami art, mostly for educational purposes.

Meenakshi Mukerji (Adhikari) was introduced to origami in early childhood by her metallurgist maternal uncle Bireshwar Mukhopadhyay. She rediscovered origami in its modular form as an adult, quite by chance in 1995, when she was living in Pittsburgh, Pennsylvania. A friend, Shobha Prabakar, took her to a class taught by Doug Philips, and ever since she has been folding modular origami and displaying it on her very popular website, http://www.origamee.net.

In 2005, Origami USA presented her with the Florence Temko award for generously sharing her modular origami work. In April 2007, A K Peters, Ltd. published her first book, *Marvelous Modular Origami*, soon to be followed by her second book, *Ornamental Origami: Exploring 3D Geometric Designs*. Both became origami best-sellers within a year of their releases.

Meenakshi was born and raised in Kolkata, India. She obtained her BS in electrical engineering at the prestigious Indian Institute of Technology, Kharagpur, and then came to the United States to pursue a master's degree in computer science at Portland State University, Oregon. After successful completion of her studies, she joined the software industry and worked for more than a decade. She is now at home in California with her husband and two sons to enrich their lives, to create her own origami designs, and to author origami books. People who have provided her with much origami inspiration and encouragement are Rosalinda Sanchez, David Petty, Francis Ow, Rona Gurkewitz, Robert Lang, Rachel Katz, Ravi Apte, and the numerous visitors of her website.

Press and Other Mentions

◈ Nov 2009: Featured artist at Pacific Coast Origami Conference 2009, San Francisco, CA.

◈ Aug 2009: Featured artist at Origami Heaven Exhibition, Stony Brook, NY.

◈ May 21, 2008: Community newspaper *The Cupertino Courier* featured a cover story on the author's origami work titled "In the Fold" by Emilie Crofton.

◈ May 21, 2008: Community newspaper *The Sunnyvale Sun* featured the same article as above.

◈ May 21, 2008: *The San Jose Mercury News* published the same article as above online.

◈ Oct 11, 2007: *Hokubei News* mentioned the author in an article entitled, "Different Ways of Origami."

◈ March 27, 2007: Community newspaper *Ukiah Daily Journal* mentioned the author in an article entitled "Her Enterprise is Folding," by Katie Mintz.

◈ June 2005: Winner of the 2005 Florence Temko Award given by Origami USA.

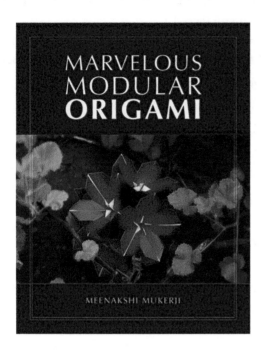

Marvelous Modular Origami,
A K Peters, Ltd., 2007.

Ornamental Origami:
Exploring 3D Geometric Designs,
A K Peters, Ltd., 2008.

Author's Contributions to Other Books and Periodicals

◈ Fifth International Conference on Origami in Science, Mathematics, and Education (5OSME) Model Collection, Singapore, July 2010; will include diagrams for Whipped Cream Star.

◈ British Origami Society, *Winchester Convention Collection, Autumn 2009,* published diagrams for Simple Cubes on CD.

◈ *Papercraft: Design and Art With Paper*, by Robert Klanten, Die Gestalten Verlag, 2009: Several of the author's creations are featured in the origami section on page 164.

◈ Origami USA magazine Paper #100, February 2009, published article titled "*An Afternoon with the Legendary Florence Temko.*"

◈ British Origami Society magazine #253, December 2008, published article titled "*An Afternoon with the Legendary Florence Temko.*"

◈ *Exhibition of Mathematical Art Catalog,* American Mathematical Society and Mathematical Association of America Joint Mathematics Meetings, San Diego, CA, 2008: Poinsettia Floral Ball photo was published and was also featured on the catalog cover.

◈ *Origami USA Annual Convention Collection, June 2008,* published diagrams for Flowered Sonobe model.

◈ British Origami Society, *Nottingham Convention, Spring 2008,* published diagrams for Blintz Base models on CD.

◈ Origami Bolivia 2008 Convention Collection published diagrams for Flower Cube model.

◈ French Origami Society MFPP Convention Collection, May 2007, published diagrams for Flowered Sonobe model.

◈ *Papiroflexia y Geometría* by Eduardo Gil Moré, published by Miguel A Salvatella, Spain, 2007: The Plain Cube unit and adaptations were published.

◈ *Ultrapassando Fronteiras com o Origami (Exceeding Borders with Origami)* by Bruno Ferraz, published by Editora Ciência Moderna, Brazil, 2007: Whirl Cube model was published.

◈ British Origami Society, *40th Anniversary Convention 2007 Collection*, published diagrams for Flowered Sonobe on CD.

◈ British Origami Society, *Bristol Convention, Autumn 2006,* published diagrams for Star Windows on CD.

◈ *The Encyclopedia of Origami* by Nick Robinson, Running Press, 2004: A full page photo of QRSTUVWXYZ Stars model appears on page 131.

◈ *Reader's Digest*, June 2004 issue, Australia Edition: A photo of QRSTUVWXYZ Stars model appears on page 17.

◈ *Reader's Digest*, June 2004 issue, New Zealand Edition: A photo of QRSTUVWXYZ Stars model appears on page 15.

◈ *Quadrato Magico #71*, August 2003 issue (a publication of the Italian Origami Society, Centro Diffusione Origami): Diagrams for Primrose Floral Ball appear on page 56.

◈ *Dave's Origami Emporium* (http://www.davidpetty.me.uk/): This website by David Petty features Planar Series diagrams in the Special Guests section (May–Aug 2003).

◈ *Scaffold*, Vol. 1 Issue 3, April 2000: Diagrams for Thatch Cube model appeared on page 4.

Author's Website

Meenakshi's Modular Mania, (http://www.origamee.net): Maintained by the author since 1997 until present, colorful and vibrant, the website features photo galleries of her own works as well as others' works folded by her. You'll also find a big collection of diagrams of some of her designs and links to diagrams of other people's designs. It is one of the top websites for "modular origami" web search and has had over a million hits. About half the models featured on her website are made with recycled or reused paper, and the author takes pride not only in that but also in that such spectacular objects can be created almost out of nothing. Immensely inspiring comments left by admirers of her website include, "This is such a great site! I was looking for some fun projects to do with kids that I tutor in math and ended up getting completely hooked myself" (Stephanie, 2006); "Your website has single handedly inspired me to explore this wonderful world of modular origami. Thank you for showing us the beautiful side of patience and perseverance" (Anonymous); "Your site shows us the beauty of color and form and that Origami is the art of possibility" (Rose J, 2005); "I've visited this website time and time again, and it never fails to take my breath away. Stunning pictures and stunning modulars" (Vigneshwaran Cumareshan, UK, 2003); "Thanks so much for all your diagrams. I'm getting extra credit in my Geometry class!" (Lawrence Date, 2001); and "...Frankly, I never knew one could create sheer poetry on paper without letting a drop of ink touch it" (Anindya Banerjee, India, 1997).